S0-CBI-385

INTERNATIONAL UNION OF PURE AND APPLIED CHEMISTRY

ANALYTICAL CHEMISTRY DIVISION
COMMISSION ON SOLUBILITY DATA

SOLUBILITY DATA SERIES

Volume 18

TETRAPHENYLBORATES

SOLUBILITY DATA SERIES

Volume 1	H. L. Clever, *Helium and Neon*
Volume 2	H. L. Clever, *Krypton, Xenon and Radon*
Volume 3	M. Salomon, *Silver Azide, Cyanide, Cyanamides, Cyanate, Selenocyanate and Thiocyanate*
Volume 4	H. L. Clever, *Argon*
Volume 5	R. Battino, *Oxygen and Ozone*
Volume 6	J. W. Lorimer, *Alkaline-earth Metal Sulfates*
Volume 7	E. M. Woolley, *Silver Halides*
Volume 8	P. Farrell, *Mono- and Disaccharides in Water*
Volume 9	R. Cohen-Adad, *Alkali Metal Chlorides*
Volume 10	J. E. Bauman, *Alkali Metal and Alkaline-earth Metal Oxides and Hydroxides in Water*
Volume 11	B. Scrosati and C. A. Vincent, *Alkali Metal, Alkaline-earth Metal and Ammonium Halides. Amide Solvents*
Volume 12	Z. Galus and C. Guminski, *Metals in Mercury*
Volume 13	C. L. Young, *Oxides of Nitrogen, Sulfur and Chlorine*
Volume 14	R. Battino, *Nitrogen*
Volume 15	H. L. Clever and W. Gerrard, *Hydrogen Halides in Non-aqueous Solvents*
Volume 16	A. L. Horvath, *Halogenated Benzenes*
Volume 17	E. Wilhelm and C. L. Young, *Hydrogen, Deuterium, Fluorine and Chlorine*
Volume 18	O. Popovych, *Tetraphenylborates*

A further 60-80 volumes are in progress to complete the Series.

SOLUBILITY DATA SERIES

Volume 18

TETRAPHENYLBORATES

Volume Editor

OREST POPOVYCH
Brooklyn College
City University of New York
USA

Evaluator and Compiler

OREST POPOVYCH

PERGAMON PRESS

OXFORD · NEW YORK · TORONTO · SYDNEY · PARIS · FRANKFURT

U.K.	Pergamon Press Ltd., Headington Hill Hall, Oxford OX3 0BW, England
U.S.A.	Pergamon Press Inc., Maxwell House, Fairview Park, Elmsford, New York 10523, U.S.A.
CANADA	Pergamon of Canada, Suite 104, 150 Consumers Road, Willowdale, Ontario M2J 1P9, Canada
AUSTRALIA	Pergamon Press (Aust.) Pty. Ltd., P.O. Box 544, Potts Point, N.S.W. 2011, Australia
FRANCE	Pergamon Press SARL, 24 rue des Ecoles, 75240 Paris, Cedex 05, France
FEDERAL REPUBLIC OF GERMANY	Pergamon Press GmbH, 6242 Kronberg-Taunus, Hammerweg 6, Federal Republic of Germany

First edition 1981

British Library Cataloguing in Publication Data

Tetraphenylborates. - (International Union of Pure and Applied Chemistry. Solubility data series; vol. 18).
1. Boron
2. Aromatic compounds
I. Popovych, Orest II. International Union of Pure and Applied Chemistry. *Commission on Solubility Data* III. Series
547'.611 QD341.B/ 80-49727

ISBN 0 08 023928 5
ISBN 0191-5622

In order to make this volume available as economically and as rapidly as possible the author's typescript has been reproduced in its original form. This method has its typographical limitations but it is hoped that they in no way distract the reader.

Printed in Great Britain by A. Wheaton & Co. Ltd., Exeter

CONTENTS

SOLUBILITY DATA SERIES

Editor-in-Chief

INTERNATIONAL UNION OF PURE AND APPLIED CHEMISTRY

IUPAC Secretariat: Bank Court Chambers, 2-3 Pound Way,
Cowley Centre, Oxford OX4 3YF, UK

FOREWORD

If the knowledge is
undigested or simply wrong,
more is not better

How to communicate and disseminate numerical data effectively in chemical science and technology has been a problem of serious and growing concern to IUPAC, the International Union of Pure and Applied Chemistry, for the last two decades. The steadily expanding volume of numerical information, the formulation of new interdisciplinary areas in which chemistry is a partner, and the links between these and existing traditional subdisciplines in chemistry, along with an increasing number of users, have been considered as urgent aspects of the information problem in general, and of the numerical data problem in particular.

Among the several numerical data projects initiated and operated by various IUPAC Commissions, the *Solubility Data Project* is probably the most ambitious one. It is concerned with preparing a comprehensive critical compilation of data on solubilities in all physical systems, of gases, liquids and solids. Both the basic and applied branches of almost all scientific disciplines require a knowledge of solubilities as a function of solvent, temperature and pressure. Solubility data are basic to the fundamental understanding of processes relevant to agronomy, biology, chemistry, geology and oceanography, medicine and pharmacology, and metallurgy and materials science. Knowledge of solubility is very frequently of great importance to such diverse practical applications as drug dosage and drug solubility in biological fluids, anesthesiology, corrosion by dissolution of metals, properties of glasses, ceramics, concretes and coatings, phase relations in the formation of minerals and alloys, the deposits of minerals and radioactive fission products from ocean waters, the composition of ground waters, and the requirements of oxygen and other gases in life support systems.

The widespread relevance of solubility data to many branches and disciplines of science, medicine, technology and engineering, and the difficulty of recovering solubility data from the literature, lead to the proliferation of published data in an ever increasing number of scientific and technical primary sources. The sheer volume of data has overcome the capacity of the classical secondary and tertiary services to respond effectively.

While the proportion of secondary services of the review article type is generally increasing due to the rapid growth of all forms of primary literature, the review articles become more limited in scope, more specialized. The disturbing phenomenon is that in some disciplines, certainly in chemistry, authors are reluctant to treat even those limited-in-scope reviews exhaustively. There is a trend to preselect the literature, sometimes under the pretext of reducing it to manageable size. The crucial problem with such preselection - as far as numerical data are concerned - is that there is no indication as to whether the material was excluded by design or by a less than thorough literature search. We are equally concerned that most current secondary sources, critical in character as they may be, give scant attention to numerical data.

On the other hand, tertiary sources - handbooks, reference books, and other tabulated and graphical compilations - as they exist today, are comprehensive but, as a rule, uncritical. They usually attempt to cover whole disciplines, thus obviously are superficial in treatment. Since they command a wide market, we believe that their service to advancement of science is at least questionable. Additionally, the change which is taking place in the generation of new and diversified numerical data, and the rate at which this is done, is not reflected in an increased third-level service. The emergence of new tertiary literature sources does not parallel the shift that has occurred in the primary literature.

With the status of current secondary and tertiary services being as briefly stated above, the innovative approach of the *Solubility Data Project* is that its compilation and critical evaluation work involve consolidation and reprocessing services when both activities are based on intellectual and scholarly reworking of information from primary sources. It comprises compact compilation, rationalization and simplification, and the fitting of isolated numerical data into a critically evaluated general framework.

The *Solubility Data Project* has developed a mechanism which involves a number of innovations in exploiting the literature fully, and which contains new elements of a more imaginative approach for transfer of reliable information from primary to secondary/tertiary sources. *The fundamental trend of the Solubility Data Project is toward integration of secondary and tertiary services with the objective of producing in-depth critical analysis and evaluation which are characteristic to secondary services, in a scope as broad as conventional tertiary services.*

Fundamental to the philosophy of the project is the recognition that the basic element of strength is the active participation of career scientists in it. Consolidating primary data, producing a truly critically evaluated set of numerical data, and synthesizing data in a meaningful relationship are demands considered worthy of the efforts of top scientists. Career scientists, who themselves contribute to science by their involvement in active scientific research, are the backbone of the project. The scholarly work is commissioned to recognized authorities, involving a process of careful selection in the best tradition of IUPAC. This selection in turn is the key to the quality of the output. These top experts are expected to view their specific topics dispassionately, paying equal attention to their own contributions and to those of their peers. They digest literature data into a coherent story by weeding out what is wrong from what is believed to be right. To fulfill this task, the evaluator must cover *all* relevant open literature. No reference is excluded by design and every effort is made to detect every relevant primary source. Poor quality or wrong data are mentioned and explicitly disqualified as such. In fact, it is only when the reliable data are presented alongside the unreliable data that proper justice can be done. The user is bound to have incomparably more confidence in a succinct evaluative commentary and a comprehensive review with a complete bibliography of both good and poor data.

It is the standard practice that any given solute-solvent system consists of two essential parts: (1) critical evaluation and recommended values, and (2) compiled data sheets.

The critical evaluation part gives the following information:

(i) a verbal text of evaluation which discusses the numerical solubility information appearing in the primary sources located in the literature. The evaluation text concerns primarily the quality of data after consideration of the purity of the materials and their characterization, the experimental method employed and the uncertainties in control of physical parameters, the reproducibility of the data, the agreement of the worker's results on accepted test systems with standard values, and finally, the fitting of data, with suitable statistical tests, to mathematical functions;

(ii) a set of recommended numerical data. Whenever possible, the set of recommended data includes weighted average and standard deviations, and a set of smoothing equations derived from the experimental data endorsed by the evaluator;

(iii) whenever relevant a graphical plot of recommended data is included.

The compilation part consists of data sheets of the best experimental data in the primary literature. Generally speaking, such independent data sheets are given only to the best and endorsed data covering the known range of experimental parameters. Data sheets based on primary sources where the data are of a lower precision are given only when no better data are available. Experimental data with a precision poorer than considered acceptable are reproduced in the form of data sheets when they are the only known data for a particular system. Such data are considered to be still suitable for some applications, and their presence in the compilation should alert researchers to areas that need more work.

The typical data sheet carries the following information:

(i) components - definition of the system - their names, formulas and *Chemical Abstracts* registry numbers;

(ii) reference to the primary source where the numerical information is reported. In cases when the primary source is a less common periodical or a report document, published though of limited availability, abstract references are also given;

(iii) experimental variables;

(iv) identification of the compiler;

(v) experimental values as they appear in the primary source. Whenever available, the data may be given both in tabular and graphical form. If auxiliary information is available, the experimental data are converted also to SI units by the compiler.

Under the general heading of auxiliary information, the essential experimental details are summarized:

(vi) experimental method used for the generation of data;
(vii) type of apparatus and procedure employed;
(viii) source and purity of materials;
(ix) estimated error;
(x) references relevant to the generation of experimental data as cited in the primary source.

This new approach to numerical data presentation has been strongly influenced by the diversity of background of those whom we are supposed to serve. We thus deemed it right to preface the evaluation/compilation sheets in each volume with a detailed discussion of the principles of the accurate determination of relevant solubility data and related thermodynamic information.

Finally, the role of education is more than corollary to the efforts we are seeking. The scientific standards advocated here are necessary to strengthen science and technology, and should be regarded as a major effort in the training and formation of the next generation of scientists and engineers. Specifically, we believe that there is going to be an impact of our project on scientific communication practices. The quality of consolidation adopted by this program offers down-to-earth guidelines, concrete examples which are bound to make primary publication services more responsive than ever before to the needs of users. The self-regulatory message to scientists of 15 years ago to refrain from unnecessary publication has not achieved much. The literature is still cluttered with poor quality articles. The Weinberg Report (in *Reader in Science Information,* Eds. J. Sherrod and A. Hodina, Microcard Editions Books, Indian Head, Inc., 1973, p. 292) states that "admonition to authors to restrain themselves from premature, unnecessary publication can have little effect unless the climate of the entire technical and scholarly community encourages restraint ..." We think that projects of this kind translate the climate into operational terms by exerting pressure on authors to avoid submitting low-grade material. The type of our output, we hope, will encourage attention to quality as authors will increasingly realize that their work will not be suited for permanent retrievability unless it meets the standards adopted in this project. It should help to dispel confusion in the minds of many authors of what represents a permanently useful bit of information of an archival value, and what does not.

If we succeed in that aim, even partially, we have then done our share in protecting the scientific community from unwanted and irrelevant, wrong numerical information.

A. S. Kertes

PREFACE

Tetraphenylborates -- compounds containing the $(C_6H_5)_4B^-$ anion, now commonly abbreviated as BPh_4^-, are a relatively new addition to chemistry, so that published material on their solubilities is, understandably, limited. Furthermore, the paucity of data on the solubility of tetraphenylborates is compounded by the fact that divalent and multivalent cations are not known to form tetraphenylborates, but instead cause the decomposition of the BPh_4^- ion. A rare exception in this respect seems to be the complex salt of ruthenium o-phenanthroline, for which solubility data are included in this Volume.

The first mention of tetraphenylborates in the literature is believed to be the 1947 report by Wittig and Keicher, describing the synthesis of lithium tetraphenylborate from triphenylboron and phenyllithium (*Naturwissenschaften*, 1947, *34*, 216). Soon thereafter, the synthesis of sodium tetraphenylborate was also accomplished in the same laboratory. (In the early literature, the tetraphenylborate ion was called simply "tetraphenylboron", to be superceded temporarily by the term "tetraphenylbor*ide* ion"). Both $LiBPh_4$ and $NaBPh_4$ are appreciably soluble in water and immediately after their synthesis attracted the attention of chemists as precipitating agents for the potassium ion, which forms a tetraphenylborate that is sparingly soluble in aqueous solutions. The fact that $KBPh_4$ is by far the least soluble potassium salt in water ($\sim 1.8 \times 10^{-4}$ mol dm^{-3} at 298 K) was probably the single most important factor that led to widespread interest in the tetraphenylborate ion and added impetus to early research on tetraphenylborate as an analytical reagent.

Soon it was discovered that rubidium, cesium, thallium(I), silver and a variety of ammonium ions formed insoluble tetraphenylborates in aqueous solution that could serve as a basis for the detection and quantitative determination of these cations. Thus, the early studies of the solubilities of tetraphenylborates were generally incidental to the development of gravimetric and other analytical procedures, where the primary focus was on the sensitivity of analysis, completeness of precipitation and thermal stability of the precipitates. Given this type of emphasis, many of the solubility data were not of the highest precision and accuracy. Furthermore, much of the analytical work, such as the spot tests for the detection of basic nitrogen compounds, was qualitative in nature, leading merely to the estimation of the orders of magnitude of the solubilities. Therefore, no compilations are provided here for published work of this type. On all aspects of the early studies on tetraphenylborates, the reader is referred to the comprehensive 1960 review by Flaschka and Barnard (in Advances in Analytical Chemistry and Instrumentation, Reilly, C. N., Ed., Interscience Publishers, Inc. New York. 1960. Vol. I).

In the 1960's and beyond, interest in tetraphenylborates was rekindled for reasons other than analytical: the BPh_4^- anion acquired the status of a "reference" ion for a variety of physico-chemical purposes. Thus, Fuoss proposed the estimation of limiting conductivities for single ions in nonaqueous solvents based on the assumption that the limiting conductivities of the BPh_4^- anion and the *n*-butyltriisopentylammonium (in the original, triisoamyl-*n*-butylammonium) cation were equal. Subsequently, we have witnessed the development and application of many analogous assumptions where a thermodynamic property of the BPh_4^- anion was equated to the corresponding property of the tetraphenylarsonium or the tetraphenylphosphonium cation. Such assumptions have been extensively applied particularly to the transfer free energies between pairs of solvents. Because values of solubility (ion-activity) products in different solvents are required for the calculation of the transfer free energies, the majority of recent data on the solubility of tetraphenylborates derives from studies where the ultimate objective is the evaluation of the transfer free energies.

Unfortunately, also here many of the published results are not of high quality. Many investigators were satisfied with results expressed to only one or two significant digits. Many have failed to specify such crucial experimental conditions as the extent of temperature control and the method of ascertaining saturation.

Tetraphenylborates are susceptible to decomposition by water, oxygen and acids. The decomposition can be easily detected by uv-spectrophotometry in the 260-280 nm range, but may pass unnoticed when a different analytical method is employed for the determination of the tetraphenylborate. This problem must be borne in mind when evaluating literature data.

Frequently, authors have reported the solubility (ion-activity) product, but not the solubility itself. In such cases, the solubility may be estimated if the correction for the activity coefficients is known or can be neglected. Thermodynamic solubility (ion-activity) products are denoted here by the symbol K°_{s0}, while the concentration solubility products are denoted by K_{s0}. Almost invariably, the solubilities in the original literature were reported in the units of molarity, here converted to the SI equivalent of mol dm^{-3}, referring to moles of the solute per dm^3 of the saturated solution. Occasionally, the reported data were in the units of grams of solute per 100 cm^3 of solution, in which case both the original results and the correspoding values in the units of mol dm^{-3} were presented in the Volume. Temperatures have been converted to Kelvin. There are very few reliable solubility data on tetraphenylborates as a function of the temperature. Only in the case of $KBPh_4$ in water was it possible to express such data by means of a smoothing equation and to calculate the standard enthalpy and entropy from it. Names recommended by Chemical Abstracts and registry numbers were used when available. Common names used in the original literature sources were given in parentheses and sometimes retained in the text.

For this Volume, an attempt was made to survey the literature through the first half of 1978. A few later publications were included when they were specifically brought to the Editor's attention. Undoubtedly, there are errors and omissions in the compilations and evaluations and the Editor will be grateful to readers who will bring these to his attention.

The following associates and members of IUPAC Commission on Solubility Data V.8. as well as other reviewers of this Volume are gratefully acknowledged for their valuable suggestions: Abraham, Chan, Chantooni, Clifford, Kertes, Khoo, Kim, Kolthoff, Loening, Lorimer, and Scrosati. Above all, my special thanks is due to Mark Salomon for his active interest and valuable advice throughout all the phases of this project, including a critical review of the manuscript.

Orest Popovych

Brooklyn, New York.

INTRODUCTION TO THE SOLUBILITY OF SOLIDS IN LIQUIDS

Nature of the Project

The Solubility Data Project (SDP) has as its aim a comprehensive search of the literature for solubilities of gases, liquids, and solids in liquids or solids. Data of suitable precision are compiled on data sheets in a uniform format. The data for each system are evaluated, and where data from different sources agree sufficiently, recommended values are proposed. The evaluation sheets, recommended values, and compiled data sheets are published on consecutive pages.

This series of volumes includes solubilities of solids of all types in liquids of all types.

Definitions

A *mixture* (1,2) describes a gaseous, liquid, or solid phase containing more than one substance, when the substances are all treated in the same way.

A *solution* (1,2) describes a liquid or solid phase containing more than one substance, when for convenience one of the substances, which is called the *solvent* and may itself be a mixture, is treated differently than the other substances, which are called *solutes*. If the sum of the mole fractions of the solutes is small compared to unity, the solution is called a *dilute solution*.

The *solubility* of a substance B is the relative proportion of B (or a substance related chemically to B) in a mixture which is saturated with respect to solid B at a specified temperature and pressure. *Saturated* implies the existence of equilibrium with respect to the processes of dissolution and precipitation; the equilibrium may be stable or metastable. The solubility of a metastable substance is usually greater than that of the corresponding stable substance. (Strictly speaking, it is the activity of the metastable substance that is greater.) Care must be taken to distinguish true metastability from supersaturation, where equilibrium does not exist.

Either point of view, mixture or solution, may be taken in describing solubility. The two points of view find their expression in the quantities used as measures of solubility and in the reference states used for definition of activities and activity coefficients.

The qualifying phrase "substance related chemically to B" requires comment. The composition of the saturated mixture (or solution) can be described in terms of any suitable set of thermodynamic components. Thus, the solubility of a salt hydrate in water is usually given as the relative proportion of anhydrous salt in solution, rather than the relative proportions of hydrated salt and water.

Quantities Used as Measures of Solubility

1. *Mole fraction* of substance B, x_B:

$$x_B = n_B / \sum_{i=1}^{c} n_i \qquad (1)$$

where n_i is the amount of substance of substance i, and c is the number of distinct substances present (often the number of thermodynamic components in the system). *Mole per cent* of B is 100 x_B.

2. *Mass fraction* of substance B, w_B:

$$w_B = m'_B / \sum_{i=1}^{c} m'_i \qquad (2)$$

where m'_i is the mass of substance i. *Mass per cent* of B is 100 w_B. The equivalent terms weight fraction and weight per cent are not used.

3. *Solute mole (mass) fraction* of solute B (3,4):

$$x_{S,B} = n_B / \sum_{i=1}^{c'} n_i = x_B / \sum_{i=1}^{c'} x_i \qquad (3)$$

where the summation is over the solutes only. For the solvent A, $x_{S,A} = x_A$. These quantities are called *Jänecke mole (mass) fractions* in many papers.

4. *Molality* of solute B (1,2) in a solvent A:

$$m_B = n_B/n_A M_A \qquad \text{SI base units: mol kg}^{-1} \qquad (4)$$

where M_A is the molar mass of the solvent.

5. *Concentration* of solute B (1,2) in a solution of volume V:

$$c_B = [B] = n_B/V \qquad \text{SI base units: mol m}^{-3} \qquad (5)$$

The terms molarity and molar are not used.
Mole and mass fractions are appropriate to either the mixture or the solution points of view. The other quantities are appropriate to the solution point of view only. In addition of these quantities, the following are useful in conversions between concentrations and other quantities.

6. *Density*: $\rho = m/V$ SI base units: kg m^{-3} (6)

7. *Relative density*: d; the ratio of the density of a mixture to the density of a reference substance under conditions which must be specified for both (1). The symbol $d^{t}_{t'}$ will be used for the density of a mixture at t°C, 1 atm divided by the density of water at t'°C, 1 atm.
Other quantities will be defined in the prefaces to individual volumes or on specific data sheets.

Thermodynamics of Solubility

The principal aims of the Solubility Data Project are the tabulation and evaluation of: (a) solubilities as defined above; (b) the nature of the saturating solid phase. Thermodynamic analysis of solubility phenomena has two aims: (a) to provide a rational basis for the construction of functions to represent solubility data; (b) to enable thermodynamic quantities to be extracted from solubility data. Both these aims are difficult to achieve in many cases because of a lack of experimental or theoretical information concerning activity coefficients. Where thermodynamic quantities can be found, they are not evaluated critically, since this task would involve critical evaluation of a large body of data that is not directly relevant to solubility. The following discussion is an outline of the principal thermodynamic relations encountered in discussions of solubility. For more extensive discussions and references, see books on thermodynamics, e.g., (5-10).

Activity Coefficients (1)

(a) *Mixtures*. The activity coefficient f_B of a substance B is given by

$$RT \ln(f_B x_B) = \mu_B - \mu_B^* \qquad (7)$$

where μ_B is the chemical potential, and μ_B^* is the chemical potential of pure B at the same temperature and pressure. For any substance B in the mixture,

$$\lim_{x_B \to 1} f_B = 1 \qquad (8)$$

(b) *Solutions*.

(i) *Solute substance, B*. The molal activity coefficient γ_B is given by

$$RT \ln(\gamma_B m_B) = \mu_B - (\mu_B - RT \ln m_B)^{\infty} \qquad (9)$$

where the superscript $^{\infty}$ indicates an infinitely dilute solution. For any solute B,

$$\gamma_B^{\infty} = 1 \qquad (10)$$

Activity coefficients y_B connected with concentration c_B, and $f_{x,B}$ (called the *rational activity coefficient*) connected with mole fraction x_B are defined in analogous ways. The relations among them are (1,9):

$$\gamma_B = x_A f_{x,B} = V_A^*(1 - \sum_s c_s) y_B \qquad (11)$$

or

$$f_{x,B} = (1 + M_A \sum_s m_s)\gamma_B = V_A{}^* y_B/V_m \quad (12)$$

or

$$y_B = (V_A + M_A \sum_s m_s V_s)\gamma_B/V_A{}^* = V_m f_{x,B}/V_A{}^* \quad (13)$$

where the summations are over all solutes, $V_A{}^*$ is the molar volume of the pure solvent, V_i is the partial molar volume of substance i, and V_m is the molar volume of the solution.

For an electrolyte solute $B \equiv C_{\nu+}A_{\nu-}$, the molal activity is replaced by (9)

$$\gamma_B m_B = \gamma_\pm{}^\nu m_B{}^\nu Q^\nu \quad (14)$$

where $\nu = \nu_+ + \nu_-$, $Q = (\nu_+{}^{\nu_+}\nu_-{}^{\nu_-})^{1/\nu}$, and $\gamma_\pm$ is the mean ionic molal activity coefficient. A similar relation holds for the concentration activity $y_B c_B$. For the mol fractional activity,

$$f_{x,B}\, x_B = \nu_+{}^{\nu_+} \nu_-{}^{\nu_-} f_\pm{}^\nu x_\pm{}^\nu \quad (15)$$

The quantities x_+ and x_- are the ionic mole fractions (9), which for a single solute are

$$x_+ = \nu_+ x_B/[1+(\nu-1)x_B]; \qquad x_- = \nu_- x_B/[1+(\nu-1)x_B] \quad (16)$$

(ii) *Solvent, A:*

The *osmotic coefficient*, ϕ, of a solvent substance A is defined as (1):

$$\phi = (\mu_A{}^* - \mu_A)/RT\, M_A \sum_s m_s \quad (17)$$

where $\mu_A{}^*$ is the chemical potential of the pure solvent.

The *rational osmotic coefficient*, ϕ_x, is defined as (1):

$$\phi_x = (\mu_A - \mu_A{}^*)/RT \ln x_A = \phi M_A \sum_s m_s/\ln(1 + M_A \sum_s m_s) \quad (18)$$

The activity, a_A, or the activity coefficient f_A is often used for the solvent rather than the osmotic coefficient. The activity coefficient is defined relative to pure A, just as for a mixture.

The Liquid Phase

A general thermodynamic differential equation which gives solubility as a function of temperature, pressure and composition can be derived. The approach is that of Kirkwood and Oppenheim (7). Consider a solid mixture containing c' thermodynamic components i. The Gibbs-Duhem equation for this mixture is:

$$\sum_{i=1}^{c'} x_i{}'(S_i{}'dT - V_i{}'dp + d\mu_i) = 0 \quad (19)$$

A liquid mixture in equilibrium with this solid phase contains c thermodynamic components i, where, usually, $c \geq c'$. The Gibbs-Duhem equation for the liquid mixture is:

$$\sum_{i=1}^{c'} x_i(S_i dT - V_i dp + d\mu_i) + \sum_{i=c'+1}^{c} x_i(S_i dT - V_i dp + d\mu_i) = 0 \quad (20)$$

Eliminate $d\mu_1$ by multiplying (19) by x_1 and (20) $x_1{}'$. After some algebra, and use of:

$$d\mu_i = \sum_{j=2}^{c} G_{ij} dx_j - S_i dT + V_i dp \quad (21)$$

where (7)

$$G_{ij} = (\partial\mu_i/\partial x_j)_{T,P,x_i \neq x_j} \quad (22)$$

it is found that

$$\sum_{i=2}^{c'} \sum_{j=2}^{c} (x_i{}' - x_i x_1'/x_1) G_{ij} dx_j - (x_1{}'/x_1) \sum_{i=c'+1}^{c} \sum_{j=2}^{c} x_i G_{ij} dx_j$$

$$= \sum_{i=1}^{c'} x_i{}'(H_i - H_i{}')dT/T - \sum_{i=1}^{c'} x_i{}'(V_i - V_i{}')dp \quad (23)$$

where

$$H_i - H_i' = T(S_i - S_i') \tag{24}$$

is the enthalpy of transfer of component i from the solid to the liquid phase, at a given temperature, pressure and composition, and H_i, S_i, V_i are the partial molar enthalpy, entropy, and volume of component i. Several special cases (all with pressure held constant) will be considered. Other cases will appear in individual evaluations.

(a) Solubility as a function of temperature.

Consider a binary solid compound A_nB in a single solvent A. There is no fundamental thermodynamic distinction between a binary compound of A and B which dissociates completely or partially on melting and a solid mixture of A and B; the binary compound can be regarded as a solid mixture of constant composition. Thus, with c = 2, c' = 1, $x_A' = n/(n+1)$, $x_B' = 1/(n+1)$, eqn (23) becomes

$$(1/x_B - n/x_A)\{1 + \left(\frac{\partial \ell n f_B}{\partial \ell n x_B}\right)_{T,P}\}dx_B = (nH_A + H_B - H_{AB}^*)\,dT/RT^2 \tag{25}$$

where the mole fractional activity coefficient has been introduced. If the mixture is a non-electrolyte, and the activity coefficients are given by the expression for a simple mixture (6):

$$RT\,\ell n\, f_B = wx_A^2 \tag{26}$$

then it can be shown that, if w is independent of temperature, eqn (25) can be integrated (cf. (5), Chap. XXIII, sect. 5). The enthalpy term becomes

$$\begin{aligned} nH_A + H_B - H_{AB}^* &= \Delta H_{AB} + n(H_A - H_A^*) + (H_B - H_B^*) \\ &= \Delta H_{AB} + w(nx_B^2 + x_A^2) \end{aligned} \tag{27}$$

where ΔH_{AB} is the enthalpy of melting and dissociation of one mole of pure solid A_nB, and H_A^*, H_B^* are the molar enthalpies of pure liquid A and B. The differential equation becomes

$$R\,d\,\ell n\{x_B(1-x_B)^n\} = -\Delta H_{AB}\,d\left(\frac{1}{T}\right) - w\,d\left(\frac{x_A^2 + nx_B^2}{T}\right) \tag{28}$$

Integration from x_B,T to $x_B = 1/(1+n)$, T = T*, the melting point of the pure binary compound, gives:

$$\begin{aligned} \ell n\{x_B(1-x_B)^n\} \simeq\ & \ell n\{\frac{n^n}{(1+n)^{n+1}}\} - \{\frac{\Delta H_{AB}^* - T^*\Delta C_p^*}{R}\}\left(\frac{1}{T} - \frac{1}{T^*}\right) \\ & + \frac{\Delta C_p^*}{R}\ell n\left(\frac{T}{T^*}\right) - \frac{w}{R}\{\frac{x_A + nx_B}{T} - \frac{n}{(n+1)T^*}\} \end{aligned} \tag{29}$$

where ΔC_p^* is the change in molar heat capacity accompanying fusion plus decomposition of the compound at temperature T*, (assumed here to be independent of temperature and composition), and ΔH_{AB}^* is the corresponding change in enthalpy at T = T*. Equation (29) has the general form

$$\ell n\{x_B(1-x_B)^n\} = A_1 + A_2/T + A_3\ell nT + A_4(x_A^2 + nx_B^2)/T \tag{30}$$

If the solid contains only component B, n = 0 in eqn (29) and (30).

If the infinite dilution standard state is used in eqn (25), eqn (26) becomes

$$RT\,\ell n\, f_{x,B} = w(x_A^2 - 1) \tag{31}$$

and (27) becomes

$$nH_A + H_B - H_{AB} = (nH_A^* + H_B^\infty - H_{AB}^*) + n(H_A - H_A^*) + (H_B - H_B^\infty) = \Delta H_{AB}^\infty + w(nx_B^2 + x_A^2 - 1) \tag{32}$$

where the first term, ΔH_{AB}^∞, is the enthalpy of melting and dissociation of solid compound A_nB to the infinitely dilute state of solute B in solvent A; H_B^∞ is the partial molar enthalpy of the solute at infinite dilution. Clearly, the integral of eqn (25) will have the same form as eqn (29), with $\Delta H_{AB}^\infty(T^*)$, $\Delta C_p^\infty(T^*)$ replacing ΔH_{AB}^* and ΔC_p^* and $x_A^2 - 1$ replacing x_A^2 in the last term.

If the liquid phase is an aqueous electrolyte solution, and the solid is a salt hydrate, the above treatment needs slight modification. Using rational mean activity coefficients, eqn (25) becomes

$$R\nu(1/x_B-n/x_A)\{1+(\partial \ln f_\pm/\partial \ln x_\pm)_{T,P}\}dx_B/\{1+(\nu-1)x_B\}$$

$$= \{\Delta H^\infty_{AB} + n(H_A-H_A^*) + (H_B-H_B^\infty)\}d(1/T) \qquad (33)$$

If the terms involving activity coefficients and partial molar enthalpies are negligible, then integration gives (cf. (11)):

$$\ln\{\frac{x_B^\nu(1-x_B)^n}{1+(\nu-1)x_B}{}^{n+\nu}\} = \ln\{\frac{n^n}{(n+\nu)^{n+\nu}}\} - \{\frac{\Delta H^\infty_{AB}(T^*)-T^*\Delta C_p^*}{R}\}(\frac{1}{T}-\frac{1}{T^*})+\frac{\Delta C_p^*}{R}\ln(T/T^*) \qquad (34)$$

A similar equation (with $\nu=2$ and without the heat capacity terms) has been used to fit solubility data for some $MOH=H_2O$ systems, where M is an alkali metal; the enthalpy values obtained agreed well with known values (11). In many cases, data on activity coefficients (9) and partial molal enthalpies (8,10) in concentrated solution indicate that the terms involving these quantities are not negligible, although they may remain roughly constant along the solubility temperature curve.

The above analysis shows clearly that a rational thermodynamic basis exists for functional representation of solubility-temperature curves in two-component systems, but may be difficult to apply because of lack of experimental or theoretical knowledge of activity coefficients and partial molar enthalpies. Other phenomena which are related ultimately to the stoichiometric activity coefficients and which complicate interpretation include ion pairing, formation of complex ions, and hydrolysis. Similar considerations hold for the variation of solubility with pressure, except that the effects are relatively smaller at the pressures used in many investigations of solubility (5).

(b) *Solubility as a function of composition.*
At constant temperature and pressure, the chemical potential of a saturating solid phase is constant:

$$\mu^*_{A_nB} = \mu_{A_nB}(\text{sln}) = n\mu_A + \mu_B \qquad (35)$$

$$= (n\mu^*_A + \nu_+\mu_+^\infty + \nu_-\mu_-^\infty) + nRT\ln f_A x_A$$

$$+ \nu RT \ln \gamma_\pm m_\pm Q_\pm \qquad (36)$$

for a salt hydrate A_nB which dissociates to water, (A), and a salt, B, one mole of which ionizes to give ν_+ cations and ν_- anions in a solution in which other substances (ionized or not) may be present. If the saturated solution is sufficiently dilute, $f_A = x_A = 1$, and the quantity K^0_{s0} in

$$\Delta G^\infty \equiv (\nu_+\mu_+^\infty + \nu_-\mu_-^\infty + n\mu^*_A - \mu_{AB}^*)$$

$$= -RT \ln K^0_{s0}$$

$$= -RT \ln Q^\nu \gamma_\pm^\nu m_+^{\nu_+} m_-^{\nu_-} \qquad (37)$$

is called the *solubility product* of the salt. (It should be noted that it is not customary to extend this definition to hydrated salts, but there is no reason why they should be excluded.) Values of the solubility product are often given on mole fraction or concentration scales. In dilute solutions, the theoretical behaviour of the activity coefficients as a function of ionic strength is often sufficiently well known that reliable extrapolations to infinite dilution can be made, and values of K^0_{s0} can be determined. In more concentrated solutions, the same problems with activity coefficients that were outlined in the section on variation of solubility with temperature still occur. If these complications do not arise, the solubility of a hydrate salt $C_{\nu_+}A_{\nu_-}\cdot nH_2O$ in the presence of other solutes is given by eqn (36) as

$$\nu \ln\{m_B/m_B(0)\} = -\nu\ln\{\gamma_\pm/\gamma_\pm(0)\} - n \ln(a_{H_2O}/a_{H_2O}(0)) \qquad (38)$$

where a_{H_2O} is the activity of water in the saturated solution, m_B is the molality of the salt in the saturated solution, and (0) indicates absence of other solutes. Similar considerations hold for non-electrolytes.

The Solid Phase

The definition of solubility permits the occurrence of a single solid phase which may be a pure anhydrous compound, a salt hydrate, a non-stoichiometric compound, or a solid mixture (or solid solution, or "mixed crystals"), and may be stable or metastable. As well, any number of solid phases consistent with the requirements of the phase rule may be present. Metastable solid phases are of widespread occurrence, and may appear as polymorphic (or allotropic) forms or crystal solvates whose rate of transition to more stable forms is very slow. Surface heterogeneity may also give rise to metastability, either when one solid precipitates on the surface of another, or if the size of the solid particles is sufficiently small that surface effects become important. In either case, the solid is not in stable equilibrium with the solution. The stability of a solid may also be affected by the atmosphere in which the system is equilibrated.

Many of these phenomena require very careful, and often prolonged, equilibration for their investigation and elimination. A very general analytical method, the "wet residues" method of Schreinemakers (12) (see a text on physical chemistry) is usually used to investigate the composition of solid phases in equilibrium with salt solutions. In principle, the same method can be used with systems of other types. Many other techniques for examination of solids, in particular X-ray, optical, and thermal analysis methods, are used in conjunction with chemical analyses (including the wet residues method).

COMPILATIONS AND EVALUATIONS

The formats for the compilations and critical evaluations have been standardized for all volumes. A brief description of the data sheets has been given in the FOREWORD; additional explanation is given below.

Guide to the Compilations

The format used for the compilations is, for the most part, self-explanatory. The details presented below are those which are not found in the FOREWORD or which are not self-evident.

Components. Each component is listed according to IUPAC name, formula, and Chemical Abstracts (CA) Registry Number. The formula is given either in terms of the IUPAC or Hill (13) system and the choice of formula is governed by what is usual for most current users: i.e. IUPAC for inorganic compounds, and Hill system for organic compounds. Components are ordered according to:

(a) saturating components;
(b) non-saturating components in alphanumerical order;
(c) solvents in alphanumerical order.

The saturating components are arranged in order according to a 18-column, 2-row periodic table:

Columns 1,2: H, groups IA, IIA;
3,12: transition elements (groups IIIB to VIIB, group VIII, groups IB, IIB);
13-18: groups IIIA-VIIA, noble gases.
Row 1: Ce to Lu;
Row 2: Th to the end of the known elements, in order of atomic number.

Salt hydrates are generally not considered to be saturating components since most solubilities are expressed in terms of the anhydrous salt. The existence of hydrates or solvates is carefully noted in the texts, and CA Registry Numbers are given where available, usually in the critical evaluation. Mineralogical names are also quoted, along with their CA Registry Numbers, again usually in the critical evaluation.

Original Measurements. References are abbreviated in the forms given by *Chemical Abstracts Service Source Index (CASSI)*. Names originally in other than Roman alphabets are given as transliterated by *Chemical Abstracts*.

Experimental Values. Data are reported in the units used in the original publication, with the exception that modern *names* for units and quantities are used; e.g., mass per cent for weight per cent; mol dm^{-3} for molar; etc. Both mass and molar values are given. Usually, only one type of value (e.g., mass per cent) is found in the original paper, and the compiler has added the other type of value (e.g., mole per cent) from computer calculations based on 1976 atomic weights (14). Errors in calculations and fitting equations in original papers have been noted and corrected, by computer calculations where necessary.

Method. Source and Purity of Materials. Abbreviations used in *Chemical Abstracts* are often used here to save space.

Estimated Error. If these data were omitted by the original authors, and if relevant information is available, the compilers have attempted to estimate errors from the internal consistency of data and type of apparatus used. Methods used by the compilers for estimating and reporting errors are based on the papers by Ku and Eisenhart (15).

Comments and/or Additional Data. Many compilations include this section which provides short comments relevant to the general nature of the work or additional experimental and thermodynamic data which are judged by the compiler to be of value to the reader.

References. See the above description for Original Measurements.

Guide to the Evaluations

The evaluator's task is to check whether the compiled data are correct, to assess the reliability and quality of the data, to estimate errors where necessary, and to recommend "best" values. The evaluation takes the form of a summary in which all the data supplied by the compiler have been critically reviewed. A brief description of the evaluation sheets is given below.

Components. See the description for the Compilations.

Evaluator. Name and date up to which the literature was checked.

Critical Evaluation

(a) Critical text. The evaluator produces text evaluating *all* the published data for each given system. Thus, in this section the evaluator review the merits or shortcomings of the various data. Only published data are considered; even published data can be considered only if the experimental data permit an assessment of reliability.

(b) Fitting equations. If the use of a smoothing equation is justifiable, the evaluator may provide an equation representing the solubility as a function of the variables reported on all the compilation sheets.

(c) Graphical summary. In addition to (b) above, graphical summaries are often given.

(d) Recommended values. Data are *recommended* if the results of at least two independent groups are available and they are in good agreement, and if the evaluator has no doubt as to the adequacy and reliability of the applied experimental and computational procedures. Data are reported as *tentative* if only one set of measurements is available, or if the evaluator considers some aspect of the computational or experimental method as mildly undesirable but estimates that it should cause only minor errors. Data are considered as *doubtful* if the evaluator considers some aspect of the computational or experimental method as undesirable but still considers the data to have some value in those instances where the order of magnitude of the solubility is needed. Data determined by an inadequate method or under ill-defined conditions are *rejected*. However references to these data are included in the evaluation together with a comment by the evaluator as to the reason for their rejection.

(e) References. All pertinent references are given here. References to those data which, by virtue of their poor precision, have been rejected and not compiled are also listed in this section.

(f) Units. While the original data may be reported in the units used by the investigators, the final recommended values are reported in S.I. units (1,16) when the data can be accurately converted.

References

1. Whiffen, D. H., ed., *Manual of Symbols and Terminology for Physicochemical Quantities and Units.* *Pure Applied Chem.* 1979, *51*, No. 1.
2. McGlashan, M.L. *Physicochemical Quantities and Units.* 2nd ed. Royal Institute of Chemistry. London. 1971.
3. Jänecke, E. *Z. Anorg. Chem.* 1906, *51*, 132.
4. Friedman, H.L. *J. Chem. Phys.* 1960, *32*, 1351.
5. Prigogine, I.; Defay, R. *Chemical Thermodynamics.* D.H. Everett, transl. Longmans, Green. London, New York, Toronto. 1954.
6. Guggenheim, E.A. *Thermodynamics.* North-Holland. Amsterdam. 1959. 4th ed.
7. Kirkwood, J.G.; Oppenheim, I. *Chemical Thermodynamics.* McGraw-Hill, New York, Toronto, London. 1961.
8. Lewis, G.N.; Randall, M. (rev. Pitzer, K.S.; Brewer, L.). *Thermodynamics.* McGraw Hill. New York, Toronto, London. 1961. 2nd ed.
9. Robinson, R.A.; Stokes, R.H. *Electrolyte Solutions.* Butterworths. London. 1959, 2nd ed.
10. Harned, H.S.; Owen, B.B. *The Physical Chemistry of Electrolytic Solutions.* Reinhold. New York. 1958. 3rd ed.
11. Cohen-Adad, R.; Saugier, M.T.; Said, J. *Rev. Chim. Miner.* 1973, *10*, 631.
12. Schreinemakers, F.A.H. *Z. Phys. Chem., stoechiom. Verwandschaftsl.* 1893, *11*, 75.
13. Hill, E.A. *J. Am. Chem. Soc.* 1900, *22*, 478.
14. IUPAC Commission on Atomic Weights. *Pure Appl. Chem.*, 1976, *47*, 75.

15. Ku, H.H., p. 73; Eisenhart, C., p. 69; in Ku, H.H., ed. *Precision Measurement and Calibration*. NBS Special Publication 300. Vol. 1. Washington. 1969.
16. *The International System of Units*. Engl. transl. approved by the BIPM of *Le Système International d'Unités*. H.M.S.O. London. 1970.

R. Cohen-Adad, Villeurbanne, France
J.W. Lorimer, London, Canada
M. Salomon, Fair Haven, New Jersey, U.S.A.

COMPONENTS:

(1) Lithium tetraphenylborate (1-); $LiC_{24}H_{20}B$; [14485-20-2]

(2) Water; H_2O; [7732-18-5]

ORIGINAL MEASUREMENTS:

Kirgintsev, A. N.; Kozitskii, V. P. *Izvest. Akad. Nauk SSSR, Khim. Ser.* 1968, 1170-2.

VARIABLES:

One temperature: 25.00°C

PREPARED BY:

Orest Popovych

EXPERIMENTAL VALUES:

The authors reported the solubility of $LiBPh_4$ in water as 39.4 mass %, where mass % was defined as grams of the salt in 100 cm^3 of the solution. This corresponds to a solubility of 1.21 mol dm^{-3} (compiler).

AUXILIARY INFORMATION

METHOD/APPARATUS/PROCEDURE:

Saturated solutions were prepared by shaking the suspensions in a constant-temperature bath for 6 hrs. Aliquots were removed through cotton plugs and weighed. The tetraphenylborate concentration was determined by precipitating $KBPh_4$ and weighing. $LiBPh_4$ recrystallized from water or acetone-water mixtures forms the solvate $LiBPh_4 \cdot 4H_2O$. Double recrystallization of the above solvate from absolute acetone yields a new solvate having the composition: $LiBPh_4 \cdot 1.5C_3H_6O \cdot 2.5H_2O$.

SOURCE AND PURITY OF MATERIALS:

$KBPh_4$ needed for the preparation of $LiBPh_4$ was synthesized in ether according to: $4C_6H_5MgBr + KBF_4 \rightarrow KB(C_6H_5)_4 + 4MgBrF$. The $KBPh_4$ was doubly recrystallized from aqu. acetone, its solution in acetone passed through an ion-exchange resin in Li form, the eluate evaporated under vacuum, the residue dissolved in a chilled chloroform-dichloroethane mixture and the $LiBPh_4$ precipitated by addition of cyclohexane.

ESTIMATED ERROR:

Precision ±0.5%

Temperature control: ±0.05°C

REFERENCES:

COMPONENTS:

(1) Lithium tetraphenylborate (1-); $LiC_{24}H_{20}B$; [14485-20-2]

(2) 2-Propanone (acetone); C_3H_6O; [67-64-1]

ORIGINAL MEASUREMENTS:

Kirgintsev, A. N.; Kozitskii, V. P. *Izvest. Akad. Nauk SSSR, Khim. Ser.* 1968, 1170-2.

VARIABLES:

One temperature: 25.00°C

PREPARED BY:

Orest Popovych

EXPERIMENTAL VALUES:

The authors reported the solubility of $LiBPh_4$ in acetone as 52.0 mass %, where mass % was defined as grams of the salt in 100 cm^3 of the solution. This corresponds to a solubility of 1.59 mol dm^{-3} (compiler).

AUXILIARY INFORMATION

METHOD/APPARATUS/PROCEDURE:

Saturated solutions were prepared by shaking the suspensions in a constant-temperature bath for 6 hrs. Aliquots were removed through cotton plugs and weighed. The tetraphenylborate concentration was determined by precipitating $KBPh_4$ and weighing. $LiBPh_4$ recrystallized from water or acetone-water mixtures forms the solvate $LiBPh_4 \cdot 4H_2O$. Double recrystallization of the above solvate from absolute acetone yields a new solvate having the composition: $LiBPh_4 \cdot 1.5C_3H_6O \cdot 2.5H_2O$.

SOURCE AND PURITY OF MATERIALS:

$KBPh_4$ needed for the preparation of $LiBPh_4$ was synthesized in ether according to: $4C_6H_5MgBr + KBF_4 \rightarrow KB(C_6H_5)_4 + 4MgBrF$. The $KBPh_4$ was doubly recrystallized from aqu. acetone and its solution in acetone passed through an ion-exchange resin in Li form. The eluate was evaporated under vacuum, the residue dissolved in a chilled chloroform-dichloroethane mixture and the $LiBPh_4$ precipitated by addition of cyclohexane. Absolute acetone was prepared by treatment with $KMnO_4$ followed by triple fractionation. Final water content was 0.007 vol. % by Karl Fisher titration.

ESTIMATED ERROR:

Precision ±0.5%

Temperature control: ±0.05°C

COMPONENTS:

(1) Sodium tetraphenylborate (1-); $NaC_{24}H_{20}B$; [143-66-8]

(2) Water; H_2O; [7732-18-5]

EVALUATOR:

Orest Popovych, Department of Chemistry, City University of New York, Brooklyn College, Brooklyn, N. Y. 11210, U. S. A.
October 1979

CRITICAL EVALUATION:

The only datum on the solubility of sodium tetraphenylborate ($NaBPh_4$) in water which is backed up by detailed and unambiguous information on the experimental conditions involved is that reported by Kirgintsev and Kozitskii (1) as part of their study of the solubility in the acetone-water system (see compilation on the solubility of $NaBPh_4$ in acetone-water mixtures). At 298.15 K, the solubility reported in the above study was 32.4 (wt./vol.)%, recalculated by the compiler to be 0.947 mol dm^{-3}. Another solubility value for $NaBPh_4$ in water can be found in the chapter by Flaschka and Barnard (2) citing as the source "personal communication" from H. Buechl. It is reported to be approximately 0.88 mol dm^{-3}, with the temperature being either 297 for 298 K. The determination was described as "direct analysis", presumably meaning the method of evaporation and weighing. However, as the experimental details related in the review of Flaschka and Barnard (2) are rather sketchy and an original report is in fact unavailable, no compilation sheet is provided for this original datum. In the case of $KBPh_4$, the solubility in water increased by about 2% per degree, so that if the same temperature dependence governs the solubility of $NaBPh_4$, one could not attribute the difference between 0.947 mol dm^{-3} and 0.88 mol dm^{-3} to the possible difference of 1 K in the temperature. A third literature value related to the solubility of $NaBPh_4$ in water is the corresponding solubility product listed as K°_{s0} = 2.14 x 10^{-2} (presumably mol^2 dm^{-6}) in the book by Clifford (3). Unfortunately, the original source of that value is not given there and taking simply the square root of the K°_{s0} leads to the value 0.146, suggesting that either a large activity correction had been taken into account, or it is based on very poor analytical work. Obviously, no original compilation sheet could be provided for that literature source. Considering that the work of Kirgintsev and Kozitskii (1) was carried out under good temperature control (±0.05°C), with the analysis performed both by the method of evaporation and weighing as well as by precipitation as $KBPh_4$ and bearing in mind the quality of other results from the same laboratory (see evaluation for KBPh in water), their solubility value can be taken as the <u>tentative value at 298.15 K</u>: 0.947 mol dm^{-3}.

REFERENCES:

1. Kirgintsev, A. N.; Kozitskii, V. P. *Izvest. Akad. Nauk SSSR, Khim. Ser.* 1968, 1170.
2. Buechl, H. cited as "personal communication" in Flaschka, H.; Barnard, A. J., Jr. *Advances in Analtical Chemistry and Instrumentation.* Reilley, C. N., Ed., Vol. 1, Chapter 1. Interscience. New York. 1960.
3. Clifford, A. F. *Inorganic Chemistry of Qualitative Analysis.* Prentice-Hall, Inc. Englewood Cliffs, N. J. 1961. p. 468.

COMPONENTS:

(1) Sodium tetraphenylborate (1-); $NaC_{24}H_{20}B$; [143-66-8]

(2) 2-Propanone (acetone); C_3H_6O; [67-64-1]

(3) Water; H_2O; [7732-18-5]

ORIGINAL MEASUREMENTS:

Kirgintsev, A. N.; Kozitskii, V. P. *Izvest. Akad. Nauk SSSR, Khim. Ser.* 1968, 1170-2.

VARIABLES:

Acetone-water composition
One temperature: 25.00°C

PREPARED BY:

Orest Popovych

EXPERIMENTAL VALUES:

The authors reported mass % of $NaBPh_4$ in the saturated solutions, defined as grams of the salt in 100 cm^3 of the solution. The solubilities have been recalculated to mol dm^{-3} by the compiler.

% Water in acetone Vol. %	Solubility of $NaBPh_4$ (Wt./vol.)%	$C/mol\ dm^{-3}$
0.007	42.8	1.25
2	45.4	1.32
4	47.8	1.40
8	52.1	1.52
12	55.9	1.63
15	58.3	1.70
20	59.4	1.74
25	60.1	1.76
30	59.7	1.74
37	58.8	1.72
45	57.8	1.69
52	56.7	1.66
60	54.9	1.60
70	52.0	1.52
80	48.4	1.41
90	42.4	1.24
100	32.4	0.947

AUXILIARY INFORMATION

METHOD/APPARATUS/PROCEDURE:

Evaporating and weighing. Saturated solutions prepared by shaking the suspensions in a constant-temperature bath for 6 hours. Aliquots removed through cotton plugs were evaporated first under an IR lamp and then dried for a week in a vacuum desiccator. Composition of the liquid phase was also checked by precipitation as $KBPh_4$. $NaBPh_4$ recovered from acetone and its aqueous mixtures containing up to 8% water contained the solvent. At higher water contents in the solvent, no crystal solvates were formed.

SOURCE AND PURITY OF MATERIALS:

Sodium tetraphenylborate of "analytical grade" obtained from the Apolda Co. (GDR) was purified by recrystallization from acetone-toluene, followed by dissolution in water and extraction with ether. The latter was removed in vacuo. The purity of the salt was no less than 99.6%. Absolute acetone was prepared by treating with $KMnO_4$ followed by triple fractionation. The final water content was 0.007 vol. %, by Karl Fisher titration.

ESTIMATED ERROR:

Precision ±0.5%

Temperature control: ±0.05°C

COMPONENTS:

(1) Sodium tetraphenylborate (1-); $NaC_{24}H_{20}B$; [143-66-8]
(2) N-Methyl-2-pyrrolidinone; C_5H_9NO; [872-50-4]

ORIGINAL MEASUREMENTS:

Virtanen, P. O. I.; Kerkelä, R. *Suom. Kemistil.* 1969, *B42*, 29-33.

VARIABLES:

Two temperatures: 25.00°C and 45.00°C

PREPARED BY:

Orest Popovych

EXPERIMENTAL VALUES:

The solubility of $NaBPh_4$ in N-methyl-2-pyrrolidinone was reported to be 1.19 mol dm^{-3} at 25°C and 1.54 mol dm^{-3} at 45°C.

The corresponding solubility product at 25°C, calculated as the square of the solubility, was reported in the form pK_{s0} = -0.15, where K_{s0} units are mol^2 dm^{-6}.

AUXILIARY INFORMATION

METHOD/APPARATUS/PROCEDURE:

The suspensions were shaken in thermostatted water-jacketed flasks for one day at 50°C, followed by one day at 25°C or 45°C, respectively. Analysis for BPh_4^- concentration in the saturated solutions was carried out by precipitating $KBPh_4$ or NH_4BPh_4 from aliquots in aqueous solution.

SOURCE AND PURITY OF MATERIALS:

N-Methyl-2-pyrrolidinone (General Aniline & Film Co.) was purified as in the literature (1).

ESTIMATED ERROR:

Not specified.
Temperature control; ±0.02°C

REFERENCES:

(1) Virtanen, P. O. I. *Suom. Kemistil.* 1966, *B39*, 257.

COMPONENTS:

(1) Sodium tetraphenylborate (1-); $NaC_{24}H_{20}B$; [i43-66-8]
(2) 1-Propanol; C_3H_8O; [71-23-8]

ORIGINAL MEASUREMENTS:

Abraham, M. H.; Danil de Namor, A.F. *J. Chem. Soc. Faraday Trans. 1*, 1978, *74*, 2101-10.

VARIABLES:

One temperature: 25°C

PREPARED BY:

Orest Popovych

EXPERIMENTAL VALUES:

The solubility of $NaBPh_4$ in 1-propanol was reported as 8.42×10^{-1} mol dm^{-3}. No further calculations were made because of solvate formation.

AUXILIARY INFORMATION

METHOD/APPARATUS/PROCEDURE:

Evaporation and weighing. Saturated solutions were prepared by shaking the suspensions for several days. The solvent contained no involatile material, but $NaBPh_4$ formed a solvate. Method of temperature control was not specified.

SOURCE AND PURITY OF MATERIALS:

The purification of the solvent was described in the literature (1). The source and purification of $NaBPh_4$ were not mentioned.

ESTIMATED ERROR:

Nothing specified.

REFERENCES:

(1) Abraham, M. H.; Danil de Namor, A. F.; Schulz, R. A. *J. Solution Chem.* 1977, *6*, 491.

COMPONENTS:	EVALUATOR:
(1) Potassium tetraphenylborate (1-); $KC_{24}H_{20}B$; [3244-41-5] (2) Water; H_2O; [7732-18-5]	Orest Popovych, Department of Chemistry, City University of New York, Brooklyn College, Brooklyn, N. Y. 11210, U. S. A. February 1979

CRITICAL EVALUATION:

A total of eleven publications dealing with the solubility of potassium tetraphenylborate ($KBPh_4$) in aqueous solutions have been reviewed. Nine of them report the solubility directly (1-9), while the remaining two (10, 11) report only the solubility product. Two studies report the solubility as a function of the temperature (7, 8) and only one deals with the variation of the solubility as a function of the temperature, ionic strength and pH (8). In one publication (9), the solubility was reported in a buffer solution, but not in pure water.

Three of the data had to be rejected outright (no compilation sheets provided). One was the datum of Raff and Brotz (1), who published an order-of-magnitude value for the solubility as being less than 10^{-4} mol dm^{-3}. The latter was estimated from the point of incipient turbidity observed visually by contacting equal volumes of a solution of KCl and a 0.1 mol dm^{-3} solution of $LiBPh_4$. Aside from the fact that the observed solubility corresponds not to pure water, but to a solution containing 0.05 mol dm^{-3} $LiBPh_4$, the entire determination was aimed at the order of magnitude of the solubility and not its precise value. The latter point has been unfortunately overlooked by subsequent investigators, who attributed more than one significant figure to the datum of Raff and Brotz (1). Another rejected datum was the solubility at 292 K reported as 1.12×10^{-4} mol dm^{-3} by Levina and Panteleeva (2), where no method was specified. Besides, the above value is very low, as compared to other literature data in that temperature range. Also rejected was the early value of the solubility product reported as 5×10^{-9} (presumably in mol^2 dm^{-6} units) at 290 K by Rüdorff and Zannier (10). No experimental details were provided there and the authors later revised the above value themselves (3).

The relative validities of the acceptable results must be assessed not so much with respect to the inherent precisions of the analytical methods employed (which are roughly comparable) as with respect to the experimental conditions, such as the time of equilibration and the attention to possible hydrolytic decomposition of the tetraphenylborate ion.

Solubility at 298 K.

The data compiled for 298 K include one determination by electrolytic conductance (3), three determinations by ultraviolet spectrophotometry (5, 6, 9), one by the method of evaporation and weighing (7), and one employing an amperometric titration (8).

The conductometric determination by Rüdorff and Zannier (3) from which the solubility is reported as 1.82×10^{-4} mol dm^{-3} is handicapped by the lack of data on the nature of the conductance apparatus employed and the precision of the temperature control as well as insufficient time of equilibration (4 hours). However, even if we assume that the experimental precision was consistent with the reported value, the latter was calculated from erroneous data. The authors used the value of 21 S cm^2 mol^{-1} for the molar conductivity of the tetraphenylborate ion and combined it with the known (unspecified) value for the λ^{∞} of the potassium ion to calculate the $\Lambda^{\infty}(KBPh_4)$. If the value $\lambda^{\infty}(K^+)$ = 73.50 S cm^2 mol^{-1} was used, the resulting $\Lambda^{\infty}(KBPh_4)$ was 94.50 S cm^2 mol^{-1}. However, if one uses the correct value for $\lambda^{\infty}(BPh_4^-)$, which is 19.69 S cm^2 mol^{-1} (12), the correct value for $\Lambda^{\infty}(KBPh_4)$ becomes 93.19 S cm^2 mol^{-1}. Presumably the authors calculated the solubility from the measured electrolytic conductivity κ and the calculated $\Lambda^{\infty}(KBPh_4)$ using the relationship $C = 1000\ \kappa/\Lambda^{\infty}$. On this basis, the correct solubility from their data would be 1.85×10^{-4} mol dm^{-3}. This value, however, is based on the <u>limiting</u> molar conductivity, which even at that low concentration applies only approximately. No data are available on the variation of the molar conductivity of $KBPh_4$ with concentration in water, but using the corresponding constants

COMPONENTS:

(1) Potassium tetraphenylborate (1-); $KC_{24}H_{20}B$; [3244-41-5]

(2) Water; H_2O; [7732-18-5]

EVALUATOR:

Orest Popovych, Department of Chemistry, City University of New York, Brooklyn College, Brooklyn, N. Y. 11210, U. S. A.
February 1979

CRITICAL EVALUATION:

for the variation of the molar conductivity of $NaBPh_4$ with concentration (12), it can be estimated that at 1.8 x 10^{-4} mol dm^{-3} the molar conductivity of $KBPh_4$ would be about 92.2. This would raise the calculated solubility further to 1.87 x 10^{-4} mol dm^{-3} .

In the opinion of this evaluator, the best method for determining tetraphenylborate concentration is ultraviolet spectrophotometry, as the spectra are very sensitive to decomposition, so that appreciable changes in the shape of the spectral bands and particularly in the ratio of the heights of the 266-nm and 274-nm peaks occur long before any yellow or brown color betrays the presence of decomposition. Thus, decomposition in the course of solubility determinations can be detected when UV-spectrophotometry is the analytical method employed, which is not true of the other analytical methods cited here.

Of the two solubility determinations in pure water by the method of uv-spectrophotometry, that by Popovych and Friedman (6) was carried out under more reliable experimental conditions, namely a temperature control to ±0.01°C in the bath from which water was circulated and an equilibration period of two weeks. The reported value is 1.74 x 10^{-4} mol dm^{-3} with a relative precision of ±1%. The second uv determination, by Pflaum and Howick (5), reports no information on the temperature control and the length of equilibration. Its further shortcoming is that the authors used the peak molar absorption coefficients ε_{max} determined in acetonitrile to analyze aqueous solutions. The acetonitrile ε_{max} values were 3.225 x 10^3 and 2.110 x 10^3 at 266 and 274 nm, respectively. For aqueous solutions, Popovych and Friedman (6) report the molar absorption coefficients as 3.25 x 10^3 and 2.06 x 10^3 at 266 and 274 nm, respectively (all molar absorption coefficients are in the units of dm^3 (cm mol)$^{-1}$). Nevertheless, the solubility reported by Pflaum and Howick (5), 1.78 x 10^{-4} mol dm^{-3}, agrees with the Popovych and Friedman (6) value, within experimental error. This agreement may be due to certain compensating errors and other mitigating factors. Thus, the solubility of $KBPh_4$ in the temperature range of 293-298 K varies only by 0.04 x 10^{-4} mol dm^{-3} per degree (7). Therefore, a temperature control to only ±0.5°C would suffice to keep the error within the limits of precision imposed by the analytical method (±1% relative). The ε_{max} value for the BPh_4^- anion used by Pflaum and Howick (5) in their calculation of the solubility was 2.5% too high at 274 nm and 1% too low at 266 nm. If the solubility value was determined as the average from the two wavelengths, what we may be seeing here is a compensation of errors.

In the uv-determination by McClure and Rechnitz (9), the solubility of $KBPh_4$ was not measured in pure water, but in a buffer solution consisting of 0.1 mol dm^{-3} tris(hydroxymethyl)aminomethane and 0.01 mol dm^{-3} acetic acid adjusted to pH 5.1 with $HClO_4$. Thus, the solubility value of 2.3 x 10^{-4} mol dm^{-3} reported by them is not comparable with the results at zero ionic strength. However, it may be compared with Siska's (8) data obtained at an ionic strength of 0.1 mol dm^{-3}, which we discuss later.

In excellent agreement with the uv-determined solubilities reported for pure water is the value obtained by evaporation of a saturated solution and weighing of the residue (7). Here, two solubility values are reported at 298 K, one resulting from a continuous equilibration at the stated temperature for 12 hours: 1.74 x 10^{-4} mol dm^{-3} (1.75 x 10^{-4} mol dm^{-3} in the original, due to an error in converting from wt %) and the other, resulting from a preliminary equilibration at 40°C for 6 hours, followed by 12 hours at 25°C: 1.79 x 10^{-4} mol dm^{-3} . This study is characterized by good temperature control (±0.05°C) and a reasonable time of equilibration.

COMPONENTS:

(1) Potassium tetraphenylborate (1-); $KC_{24}H_{20}B$; [3244-41-5]

(2) Water; H_2O; [7732-18-5]

EVALUATOR:

Orest Popovych, Department of Chemistry, City University of New York, Brooklyn College, Brooklyn, N. Y. 11210, U. S. A.
February 1979

CRITICAL EVALUATION:

Recommended Values at 298.15 K

Combining the average value reported by Kozitskii (7), 1.76×10^{-4} mol dm^{-3}, by Popovych and Friedman (6), 1.74×10^{-4} mol dm^{-3}, and by Pflaum and Howick (5), 1.78×10^{-4} mol dm^{-3}, we obtain the overall average from three studies by two different methods as:

$$\text{Solubility} = (1.76 \pm 0.02) \times 10^{-4} \text{ mol dm}^{-3}$$

The absolute error derives from the relative precision of a uv-determination of the tetraphenylborate concentration, which is ±1% (6).

The solubility product can be calculated as $K^\circ_{s0} = (Cy_\pm)^2$, where the mean molar ionic activity coefficient $y_\pm{}^2$ is estimated from the Debye-Hückel limiting law: $\log y_\pm{}^2 = -1.18(1.76 \times 10^{-4})^{\frac{1}{2}}$. This yields $y_\pm{}^2 = 0.969$ and

$$K^\circ_{s0} = (3.00 \pm 0.04) \times 10^{-8} \text{ mol}^2 \text{ dm}^{-6}$$

The absolute error in the K°_{s0} was calculated assuming the 1% error in C as its only source.

Solubility at Other Temperatures

At 293 K, the solubility can be found in two literature sources (4,7), reported directly and in another source where it is reported in the form of the solubility product (11). Closest agreement exists between the value published by Kozitskii (7), 1.56×10^{-4} mol dm^{-3}, which was determined by evaporation and weighing, and the radiometrically determined value of 1.48×10^{-4} mol dm^{-3} reported by Geilmann and Gebauhr (4). Unfortunately, even these two values are not quite close enough to merit averaging for a recommended value. Considering that Kozitskii specified an equilibration time of 24 hours and that his solubility at 298 K is equal to the recommended value at that temperature, this evaluator chooses 1.56×10^{-4} mol dm^{-3} as the tentative value for the solubility at 293 K. Geilman and Gebauhr (4), on the other hand, did not specify the time of equilibration.

The other solubility values reported for 293 K appear to be too low. From a potentiometric titration, Havíř (11) reports a solubility product of 1.6×10^{-8} (presumably in mol^2 dm^{-6} units), from which the solubility taken simply as the square root would be 1.3×10^{-4} mol dm^{-3}. No equilibration time was specified here, but considering that the solubility of $AgBPh_4$ reported in the same article was determined after only 4 hours of equilibration, it is likely that a similar time was used for $KBPh_4$. It should be noted that the intention of the author in this case was to demonstrate the concentration limit to which a potentiometric titration of the BPh_4^- ion was feasible and that the determination of the solubility of $KBPh_4$ was incidental. Similarly, Siska's (8) value of 1.25×10^{-4} mol dm^{-3}, which was determined after only 3 hours of equilibration, must be too low due to absence of saturation.

Both Siska (8) and Kozitskii (7) reported the solubility at several temperatures. Siska also showed tables of the solubility as a function of ionic strength and the pH. Most of his values were determined at an ionic strength of 0.1 mol dm^{-3} and are therefore not directly comparable with other literature data. However, judging from the one datum given for zero ionic strength at 293 K (1.25×10^{-4} mol dm^{-3}) and from the fact that Siska's data at other temperatures at 0.1 ionic strength are of approximately the same magnitude as other literature data at zero ionic strength, it seems that all the results in this most comprehensive study of the solubility of $KBPh_4$ in aqueous solutions are too low due to

COMPONENTS:

(1) Potassium tetraphenylborate (1-); $KC_{24}H_{20}B$; [3244-41-5]

(2) Water; H_2O; [7732-18-5]

EVALUATOR:

Orest Popovych, Department of Chemistry, City University of New York, Brooklyn College, Brooklyn, N. Y. 11210, U. S. A.
February 1979

CRITICAL EVALUATION:

undersaturation. This conclusion seems also to be corroborated by the fact that the solubility of $KBPh_4$ in a 0.1 mol dm^{-3} buffer solution reported by McClure and Rechnitz (9) at 298 K, (2.3×10^{-4} mol dm^{-3}) is higher than the solubility reported by Siska at 303 K (2.13×10^{-4} mol dm^{-3}). Consequently, we are again limited to stating as <u>tentative values</u> those solubilities that were reported by Kozitskii (7) at other temperatures as well. Below we tabulate these solubilities as well as the solubility products derived from them using mean molar activity coefficients $y_\pm$ estimated from the Debye-Hückel limiting law. It should be noted that in the concentration range involved, the use of a Debye-Hückel expression with ion-sized parameters would result in a change of about one part per thousand only.

Solubility at Different Temperatures (8)

T/K	$10^4 C$/mol dm^{-3}	$10^8 K^\circ_{s0}$/mol^2 dm^{-6}*	A (Debye-Hückel)
273.15	1.29	1.62	0.490
293.15	1.56	2.36	0.505
298.15	1.76	3.00	0.509
323.15	3.71	13.1	0.537

*Calculated by the evaluator as $K^\circ_{s0} = C^2 y_\pm^2$, where $\log y_\pm^2 = -2A(C)^{\frac{1}{2}}$ and the units of A are $mol^{-1/2}$ $dm^{3/2}$.

A plot of $\log K^\circ_{s0}$ vs. T^{-1} is linear only in the range of 293-323 K, for which the smoothing equation obtained by the method of least squares is: $\log K^\circ_{s0} = -2387/(T/K) + 0.501$, with $\sigma_y =$ 0.013 (absolute) and a correlation coefficient of -0.999. The highly tentative values for the thermodynamic constants calculated from the above slope and intercept are:

$$\Delta H^\circ + 45.7 \pm 1.8 \text{ kJ mol}^{-1} \text{ and } \Delta S^\circ = 9.6 \pm 6.1 \text{ JK}^{-1} \text{ mol}^{-1}.$$

All four points can be described by the equation:

$$\log K^\circ_{s0} = 30.97 - 2.121 - 10^4/(T/K) + 2.903 \times 10^6/(T/K)^2 \text{ with}$$
$\sigma_y =$ 0.006 (abs.).

References:

1. Raff, P.; Brotz, W. *Z. anal. Chem.* 1951, *133*, 241.
2. Levina, N. D.; Panteleeva, N. I. *Zavod. Lab.* 1957, *23*, 285.
3. Rüdorff, W.; Zannier, H. *Z. Naturforsch.* 1953, *8b*, 611.
4. Geilmann, W.; Gebauhr, W. *Z. anal. Chem.* 1953, *139*, 161.
5. Pflaum, R. T.; Howick, L. C. *Anal. Chem.* 1956, *28*, 1542.
6. Popovych, O.; Friedman, R. M. *J. Phys. Chem.* 1966, *70*, 1671.
7. Kozitskii, V. P. *Izvest. Akad. Nauk SSSR, Khim. Ser.* 1970, 8.
8. Siska, E. *Magy. Kem. Foly.* 1976, *82*, 275.
9. McClure, J. E.; Rechnitz, G. A. *Anal. Chem.* 1966, *38*, 136.
10. Rüdorff, W.; Zannier, H. *Angew. Chem.* 1952, *64*, 613.
11. Havíř, J. *Collect. Czech. Chem. Commun.* 1959, *24*, 1955.
12. Skinner, J. F.; Fuoss, R. M. *J. Phys. Chem.* 1964, *68*, 1882.

COMPONENTS:

(1) Potassium tetraphenylborate (1-); $KC_{24}H_{20}B$; [3244-41-5]

(2) Water; H_2O; [7732-18-5]

ORIGINAL MEASUREMENTS:

Rüdorff, W.; Zannier, H.

Z. Naturforsch. 1953, *8b*, 611-2.

VARIABLES:

One temperature: 25°C

PREPARED BY:

Orest Popovych

EXPERIMENTAL VALUES: The solubility of potassium tetraphenylborate ($KBPh_4$) was reported as 1.82×10^{-4} mol dm^{-3} and the solubility product, K_{s0}, as $3.3 \times 10^{-8} mol^2 dm^{-6}$.

The authors determined the limiting molar conductivity of sodium tetraphenylborate to be 71 S cm^2 mol^{-1}, from which they derived the value of 21 S cm^2 mol^{-1} for the λ^∞ of the tetraphenylborate anion. Combining the latter with the known (unspecified) value for the $\lambda^\infty(K^+)$, the authors calculated the solubility from the conductance of the saturated solution of potassium tetraphenylborate, presumably using the relationship: $C = 1000\ \kappa/\Lambda^\infty$, where C is the solubility and κ, the electrolytic conductivity of the solution corrected for solvent conductance (magnitude not specified). The calculation method, however, was not explained in the article.

The present results were reported as an improvement over the previously published by the same author (1) value of $K_{s0} = 5 \times 10^{-9}$ (presumably in units of mol^2 dm^{-6}), which was also derived from conductance data.

AUXILIARY INFORMATION

METHOD/APPARATUS/PROCEDURE:

Electrolytic conductance on unspecified apparatus. Temperature controlled, but within unspecified limits. Saturated solutions of $KBPh_4$ were prepared by bubbling nitrogen through its suspensions in conductivity water for 4 hours.

SOURCE AND PURITY OF MATERIALS:

$NaBPh_4$ from the Heyl Co. of Hildesheim, Germany, was purified in the absence of air by just dissolving it in chloroform-acetone mixture and precipitating with petroleum ether. After repeated treatment, it was recrystallized under N_2 from chloroform and vacuum dried. $KBPh_4$ was recrystallized from acetone-ethyl acetate.

ESTIMATED ERROR:

Nothing specified.

REFERENCES:

Rüdorff, W.; Zannier, H.

Angew. Chem. 1952, *64*, 613.

COMPONENTS:

(1) Potassium tetraphenylborate (1-); $KC_{24}H_{20}B$; [3244-41-5]

(2) Water; H_2O; [7732-18-5]

ORIGINAL MEASUREMENTS:

Geilmann, W.; Gebauhr, W.

Z. anal. Chem. 1953, *139*, 161-81.

VARIABLES:

One temperature: 20°C

PREPARED BY:

Orest Popovych

EXPERIMENTAL VALUES:

The solubility is reported both as C_K = 0.578 mg/100 ml of water and as 1.5 x 10^{-4} mol dm^{-3}. Retaining three significant figures, the above solubility would be 1.48 x 10^{-4} mol dm^{-3} (compiler).

The solubility product of $KBPh_4$ is reported as K_{s0} = 2.25 x 10^{-8} mol^2 dm^{-6}. Because the latter is simply $C_K{}^2$, the value calculated from the solubility expressed to three significant figures is K_{s0} = 2.19 x 10^{-8} mol^2 dm^{-6} (compiler).

Also reported is the rate of dissolution of potassium tetraphenylborate in water:

Time, hours	μg K/10 ml of water
0.5	55.7
1.0	56.0
2.0	57.0
5.0	57.4
8.0	57.8
16.0	57.5

AUXILIARY INFORMATION

METHOD/APPARATUS/PROCEDURE: Radiometric, using liquid-scintillation counting of ^{42}K. Apparatus not specified. The radioactive potassium obtained as the carbonate from the Harwell nuclear reactor was purified by precipitation with $HClO_4$ in the presence of 2 mg of Na_2HPO_4 and recrystallization. The $KClO_4$ solution was reacted with $NaBPh_4$, the resulting $KBPh_4$ precipitate washed with water, mechanically shaken in water at 20°C and the filtrate analyzed radiometrically to constant activity.

SOURCE AND PURITY OF MATERIALS:

Nothing specified.

ESTIMATED ERROR:

Not specified. However, given the temperature control to ±0.5°C, the relative precision cannot be better than ±1-2% (compiler).

REFERENCES:

COMPONENTS:

(1) Potassium tetraphenylborate (1-); $KC_{24}H_{20}B$; [3244-41-5]

(2) Water; H_2O; [7732-18-5]

ORIGINAL MEASUREMENTS:

Pflaum, R. T.; Howick, L. C.

Anal. Chem. 1956, *28*, 1542-44.

VARIABLES:

One temperature: 25°C

PREPARED BY:

Orest Popovych

EXPERIMENTAL VALUES:

The solubility of potassium tetraphenylborate ($KBPh_4$) was reported as 1.78×10^{-4} mol dm^{-3}.

Also reported were the molar absorption coefficients of the tetraphenylborate ion in acetonitrile at 266 nm and 274 nm as 3225 and 2100 dm^3 (cm mol)$^{-1}$, respectively.

AUXILIARY INFORMATION

METHOD/APPARATUS/PROCEDURE: Ultraviolet spectrophotometry on a Cary Model 11 recording spectrophotometer. Saturated solutions were prepared in conductivity water by an unspecified procedure. Method of controlling the temperature was not stated. The concentration of BPh_4^- in saturated solutions was obtained from spectrophotometric measurements at 266 and 274 nm by applying the molar absorption coefficients specified above.

SOURCE AND PURITY OF MATERIALS:

$NaBPh_4$ (J. T. Baker Chemical Co.) was used as received for pptns, but was recrystallized from acetone-hexane mixt for detn of absorption coefficients. $KBPh_4$ was prepd by metathesis of KCl and $NaBPh_4$ and purified by recrystallization from a CH_3CN-H_2O mixt. CH_3CN (Matheson, Coleman & Bell) was treated with cold satd KOH, dried over anhydrous K_2CO_3 for 24 hrs., refluxed over P_2O_5 in an all-glass apparatus. The fraction boiling at 81-81.5 °C was retained. All other chemicals were of reagent grade quality.

ESTIMATED ERROR:

Nothing specified. Precision is likely to be of the order of ±1% (compiler).

COMPONENTS:

(1) Potassium tetraphenylborate (1-); $KC_{24}H_{20}B$; [3244-41-5]

(2) Water; H_2O; [7732-18-5]

ORIGINAL MEASUREMENTS:

Popovych, O.; Friedman, R. M.

J. Phys. Chem. 1966, *70*, 1671-3.

VARIABLES:

One temperature: 25.00°C

PREPARED BY:

Orest Popovych

EXPERIMENTAL VALUES:

The solubility of potassium tetraphenylborate ($KBPh_4$) was reported as the concentration of total tetraphenylborate in its saturated solution:

$$C = 1.74 \times 10^{-4} \text{ mol dm}^{-3}$$

The solubility product $K^\circ_{s0} = (Cy_\pm)^2$ was computed by the authors from the above solubility C and the activity coefficient $y_\pm$ calculated from the Debye-Hückel equation in the form:

$$\log y_\pm = \frac{-0.509\ C^{1/2}}{1 + 0.328\mathring{a}C^{1/2}}$$

Values of the ion-size parameter $\mathring{a}$ were taken as 0.3 nm for the K^+ ion (1) and 1.0 nm for the tetraphenylborate ion (2). The solubility product computed in this manner was:

$$K^\circ_{s0} = 2.94 \times 10^{-8} \text{ mol}^2 \text{ dm}^{-6}$$

Complete dissociation was assumed.
The molar absorption coefficient of the tetraphenylborate ion in water was reported as 3.25×10^3 at 266 nm and as 2.06×10^3 at 274 nm. The units of the absorption coefficient were $dm^3(cm\ mol)^{-1}$.

AUXILIARY INFORMATION

METHOD/APPARATUS/PROCEDURE: Ultraviolet spectrophotometry using a Cary Model 14 recording spectrophotometer. Saturated solutions were prepared by shaking suspensions of $KBPh_4$ in water on a Burrell wrist-action shaker in water-jacketed flasks with water circulated from a constant-temperature bath maintained at 25.00 ±0.01°C. After about 2 weeks of shaking, the suspensions were filtered and the filtrates analyzed spectrophotometrically. The molar absorption coefficients stated above were used to compute the concentration of tetraphenylborate. All work was carried out in deaerated containers and solvents.

SOURCE AND PURITY OF MATERIALS:

$KBPh_4$ was prepared from $NaBPh_4$ (Fisher, 99.7%) by metathesis with KCl; it was recrystallized 3 times from 3:1 acetone-water and dried in vacuo at 80°C.

Deionized water was redistilled.

ESTIMATED ERROR:

Not stated by the authors, but relative precision in known to be of the order of 1% (compiler).
Temperature: ±0.01°C

REFERENCES:

(1) Kielland J. *J. Am. Chem. Soc.* 1937, *59*, 1675.

(2) Nightingale, E. R. *J. Phys. Chem.* 1959, *63*, 1381.

COMPONENTS:

(1) Potassium tetraphenylborate (1-); $KC_{24}H_{20}B$; [3244-41-5]

(2) 2-Propanone (acetone); C_3H_6O; [67-64-1]

(3) Water; H_2O; [7732-18-5]

ORIGINAL MEASUREMENTS:

Kozitskii, V. P. *Izvest. Akad. Nauk SSSR, Khim. Ser.* 1970, 8-11.

VARIABLES:

Acetone-water composition
2 temperatures: 0.00°C and 50.00°C for acetone-water mixtures and a range of 0.00-97.50°C for water.

PREPARED BY:

Orest Popovych

EXPERIMENTAL VALUES:

The author reports the solubility of potassium tetraphenylborate ($KBPh_4$) in water in the units of mg/l and in mol dm^{-3}. In acetone-water mixtures, the solubility is reported as mass % of the salt in saturated solutions. The mass % is defined as the number of grams of the salt in 100 ml of the solution. The latter solubilities have been recalculated to the units of mol dm^{-3} by the compiler.

Solubility of $KBPh_4$ in water

t/°C	mg dm^{-3}	10^4 mol dm^{-3}	hydrolysis products, mg dm^{-3}	Time of saturation, hrs	pH
0	46.1	1.29	0	24	6.6
20	56.0	1.56	0	24	6.5
25	62.5	1.75†	0	12	6.5
25*	64.0	1.79	0	18	--
50	133	3.7 x	0	8	6.35
75	301	8.1	14.4	1.1	7.40
	282		23.2	6	7.82
97.5	426	13	238	0.6	7.90
	546		94	1.5	8.85

*Preliminary equilibration at 40°C for 6 hours.
†Should be 1.74 (compiler).
xShould be 3.71 (compiler).

The above table also illustrates the extent of hydrolytic decomposition of the tetraphenylborate anion as a function of the temperature and the time of equilibration.

AUXILIARY INFORMATION

METHOD/APPARATUS/PROCEDURE:

Evaporation and weighing.
Saturated solutions were prepared by stirring mechanically the suspensions in thermostatted baths for the length of time indicated in the above Table. One liter of the filtrate from the saturated solution was dried by vacuum distillation (∿5 hrs keeping the temperature no higher than 40°C). The residue and the column were rinsed out many times with small portions of acetone, which was then collected, evaporated, and the residue, weighed.

SOURCE AND PURITY OF MATERIALS:

Absolute acetone (0.0065 vol % H_2O) was prepd by the same method as described on the compilation sheet for $KBPh_4$ in acetone-water at 25°C. $KBPh_4$ obtained by metathesis of KCl and $NaBPh_4$ was purified by double recrystallization from 3:1 acetone-water, evaporation of the acetone, washing of the crystals with water and ether and vacuum drying at 60°C. Water was doubly distilled.

ESTIMATED ERROR:

None stated.
Temperature: ±0.05°C

REFERENCES:

COMPONENTS:

(1) Potassium tetraphenylborate (1-); $KC_{24}H_{20}B$; [3244-41-5]

(2) 2-Propanone (acetone); C_3H_6O; [67-64-1]

(3) Water: H_2O; [7732-18-5]

ORIGINAL MEASUREMENTS: (continued)

Kozitskii, V. P. *Izvest. Akad. Nauk SSSR, Khim. Ser.* 1970, 8-11.

COMMENTS AND/OR ADDITIONAL DATA:

The author reports the solubility of potassium tetraphenylborate in acetone-water mixtures at 0, 25 and 50°C, but the values at 25°C were taken from a previous study in the same laboratory (1) compiled on the sheet for $KBPh_4$ in acetone-water mixtures at 25°C, with the exception of the value for pure water, which is now reported as 0.0063 wt %, corresponding to 1.76 x 10^{-4} mol dm^{-3} (compiler).

Solubility of $KBPh_4$ in Acetone-Water Mixtures

Temperature	0°C		50°C	
Vol.% H_2O in acetone	(wt/vol)%	mol dm^{-3}(compiler)	(wt/vol)%	mol dm^{-3}(compiler)
0.007	7.04	0.196	5.15	0.144
2	7.59	0.212	6.04	0.169
4	7.67	0.214	6.58	0.184
8	7.18	0.200	7.14	0.199
12	6.45	0.180	7.05	0.197
15	5.87	0.164	6.77	0.189
20	4.88	0.136	6.24	0.174
25	3.81	0.106	5.56	0.155
33.3	2.19	6.11×10^{-2}	4.24	0.118
37	---	-----	3.69	0.103
40	1.37	3.82×10^{-2}	----	-----
45	0.86	$2.4_0 \times 10^{-2}$	2.42	6.75×10^{-2}
52	0.389	1.09×10^{-2}	1.50	4.19×10^{-2}
60	0.116	3.24×10^{-3}	0.82	2.29×10^{-2}
70	0.031	8.6×10^{-4}	0.26	7.3×10^{-3}
80	0.0096	$2.6_8 \times 10^{-4}$	0.0992	2.77×10^{-3}
100	0.0046	1.3×10^{-4}	0.0133	3.71×10^{-4}

REFERENCES:

(1) Kirgintsev, A. N. Kozitskii, V. P. *Izvest. Akad. Nauk SSSR, Khim. Ser.* 1968, 1170.

COMPONENTS:

(1) Potassium tetraphenylborate (1-); $KC_{24}H_{20}B$; [3244-41-5]
(2) Sodium sulfate; Na_2SO_4; [7757-82-6]
(3) Water; H_2O; [7732-18-5]

ORIGINAL MEASUREMENTS:

Siska, E. *Magy. Kem. Foly.* 1976, *82*, 275-8.

VARIABLES:

Temperature range 10-45°C
Concentration of Na_2SO_4
pH

PREPARED BY:

Orest Popovych

EXPERIMENTAL VALUES:

The solubility of potassium tetraphenylborate ($KBPh_4$) at 20 ± 1°C was reported for distilled water to be C = 1.35 x 10^{-4} mol dm^{-3} and the corresponding solubility product, calculated at $K_{s0} = C^2$, was reported as K_{s0} = 1.72 x 10^{-8} mol^2 dm^{-6}. With ionic strength varied by means of Na_2SO_4, the following solubilities C were reported for potassium tetraphenylborate in aqueous solutions at 20 ± 1°C:

Ionic strength, mol dm^{-3}	10^4C/mol dm^{-3}
0	1.25
0.05	1.49
0.1	1.50
0.3	1.34
0.5	1.25
0.7	1.11
1.0	0.79
2.0	0.43

AUXILIARY INFORMATION

METHOD/APPARATUS/PROCEDURE:

Amperometric titration of the tetraphenylborate ion with $AgNO_3$ or $TlNO_3$ solutions using a Radeikis OH-102 polarograph and a graphite-calomel electrode system with an agar-agar bridge. The calomel electrode contained 0.1 mol dm^{-3} NaCl. The titration was carried out in 5-10 cm^3 of acetic acid-sodium acetate buffer mixed with 5 cm^3 of acetone. $KBPh_4$ was prepared by metathesis of KCl and $NaBPh_4$. Saturated solutions of $KBPh_4$ were prepared by magnetically stirring its suspensions for up to 3 hours.

SOURCE AND PURITY OF MATERIALS:

Not specified.

ESTIMATED ERROR:

±2% is the precision in the solubility determination.
Temperature ±1°C

REFERENCES:

COMPONENTS:	ORIGINAL MEASUREMENTS: (continued)
(1) Potassium tetraphenylborate (1-); $KC_{24}H_{20}B$; [3244-41-5] (2) Sodium sulfate; Na_2SO_4; [7757-82-6] (3) Water; H_2O; [7732-18-5]	Siska, E. *Magy. Kem. Foly.* 1976, *82*, 275-8.

COMMENTS AND/OR ADDITIONAL DATA:

Keeping the ionic strength constant at 0.1 mol dm^{-3} with sodium sulfate, the following solubilities C were obtained as a function of the temperature:

t/°C	10^4C/mol dm^{-3}
10	1.19
20	1.50
30	2.13
40	2.45
45	2.98

Keeping the ionic strength constant at 0.1 mol dm^{-3}, with sodium sulfate, the following solubilities C were obtained as a function of the pH varied by means of acetic acid and sodium hydroxide at 20 ± 1°C:

pH	10^4C/mol dm^{-3}	pH	10^4C/mol dm^{-3}
1.3	0.30	11.5	1.40
1.8	0.94	11.6	1.40
2.8	1.54	11.6	1.42
3.0	1.42	11.7	1.50
4.6	1.48	3.7	1.42
5.7	1.44	3.9	1.40
6.6	1.45		
6.9	1.44		
7.8	1.42		
8.8	1.42		
10.2	1.56		
10.4	1.44		
10.6	1.50		
11.2	1.46		

The authors report at an ionic strength of 0.1 and the pH range of 2.8-11.7, the following solubility value for potassium tetraphenylborate in aqueous solution:
$C = 1.45 \times 10^{-4}$ mol dm^{-3}.
The error in the above value was stated as 0.046 (10^{-4} mol dm^{-3}), presumably referring to the precision.

COMPONENTS:

(1) Potassium tetraphenylborate (1-); $KC_{24}H_{20}B$; [3244-41-5]
(2) Tris(hydroxymethyl)aminomethane; $C_4H_{11}NO_3$; [77-86-1]
(3) Acetic acid; $C_2H_4O_2$; [64-19-7]
(4) Water; H_2O; [7732-18-5]

ORIGINAL MEASUREMENTS:

McClure, J. E.; Rechnitz, G. A. *Anal. Chem.* 1966, *38*, 136-139.

VARIABLES:

One temperature: 24.8°C

PREPARED BY:

Orest Popovych

EXPERIMENTAL VALUES:

The solubility of potassium tetraphenylborate ($KBPh_4$) in aqueous tris(hydroxymethyl)aminomethane (THAM) buffer at a pH 5.1 was reported as 2.3×10^{-4} mol dm^{-3}.

AUXILIARY INFORMATION

METHOD/APPARATUS/PROCEDURE:

UV-spectrophotometry according to the procedure of Howick and Pflaum (1). No other details.

SOURCE AND PURITY OF MATERIALS:

The buffer solution was composed of 0.1 mol dm^{-3} THAM and 0.01 mol dm^{-3} acetic acid, adjusted to pH 5.1 with G. F. Smith reagent-grade $HClO_4$. The source of BPh_4^- was a solution of $Ca(BPh_4)_2$ in THAM prepared from Fisher Scientific reagent-grade $NaBPh_4$ by the procedure of Rechnitz et al. (2) and standardized by potentiometric titrn with KCl and RbCl. Baker reagent-grade KCl was the source of K^+.

ESTIMATED ERROR:

Not stated.

Temperature: ±0.3°C

REFERENCES:

1. Howick, L. C.; Pflaum, R. T. *Anal. Chim. Acta* 1958, *19*, 342.
2. Rechnitz, G. A. ; Katz, S. A.; Zamochnick, S. B. *Anal. Chem.* 1963, *35*, 1322.

COMPONENTS:

(1) Potassium tetraphenylborate (1-); $KC_{24}H_{20}B$; [3244-41-5]

(2) Water; H_2O; [7732-18-5]

ORIGINAL MEASUREMENTS:

Havíř, J. *Collect. Czech. Chem. Commun.* 1959, *24*, 1955-9.

VARIABLES:

One temperature: 20°C

PREPARED BY:

Orest Popovych

EXPERIMENTAL VALUES:

The solubility product of potassium tetraphenylborate ($KBPh_4$) in water is reported to be:
$K_{s0} = 1.6 \times 10^{-8}$ presumably in $mol^2\ dm^{-6}$ units (compiler). The corresponding solubility calculated as $(K_{s0})^{1/2}$ is:
$C = 1.3 \times 10^{-4}\ mol\ dm^{-3}$ (compiler).

AUXILIARY INFORMATION

METHOD/APPARATUS/PROCEDURE: Potentiometric titration of the tetraphenylborate ion in saturated $KBPh_4$ solutions with $AgNO_3$ using a freshly plated silver indicator electrode and a saturated calomel reference electrode dipping in a solution of 10% $NaNO_3$. The salt bridge was a 2% Agar-Agar solution in a 10% $NaNO_3$ solution. An Ionoskop potentiometer was used.

SOURCE AND PURITY OF MATERIALS:

$NaBPh_4$ was obtained from the Heyl Co. (Berlin), but the method of preparation and purification of the potassium salt was not specified.

ESTIMATED ERROR:

Nothing specified.

REFERENCES:

COMPONENTS:	EVALUATOR:
(1) Potassium tetraphenylborate (1-); $KC_{24}H_{20}B$; [3244-41-5] (2) 2-Propanone (acetone); C_3H_6O; [67-64-1] (3) Water; H_2O; [7732-18-5]	Orest Popovych, Department of Chemistry, City University of New York, Brooklyn College, Brooklyn, N. Y. 11210, U. S. A. September 1979

CRITICAL EVALUATION:

The solubility of potassium tetraphenylborate ($KBPh_4$) in acetone-water mixtures was determined at 301 K by Scott et al. (1) and at 298.15 K by Kirgintsev and Kozitskii (2). This difference in the temperature and, to a smaller degree, the fact that the acetone in the first study was of reagent grade and may have contained up to 0.5% water preclude a comparison of the two sets of data. Only for pure acetone was it possible to make a comparison between the results obtained from the above two studies, but the agreement was very poor. Thus, according to Scott et al. (1) the solubility in acetone at 301 K was 0.117 mol dm^{-3}, while a smoothing equation based on the data of Kirgintsev and Kozitskii (2, 3) predicted a solubility of 0.165 mol dm^{-3} at 301 K (see evaluation for $KBPh_4$ in acetone). Of course, it is impossible to tell whether or not the two sets of data would similarly disagree at other solvent compositions, since data for the solubility as a function of the temperature are not available for acetone-water mixtures. At this time, the solubilities reported for acetone-water mixtures by Kirgintsev and Kozitskii (2) should be regarded as the tentative values at 298.15 K. They were obtained under good temperature control (±0.05°C), but the equilibration time of 6 hours may have been insufficient for complete saturation.

REFERENCES:

1. Scott, A. D.; Hunziker, H. H.; Reed, M. G. *Chemist-Analyst* 1959, *48*, 11.
2. Kirgintsev, A. N.; Kozitskii, V. P. *Izvest. Akad. Nauk SSSR, Khim. Ser.* 1968, 1170.
3. Kozitskii, V. P. *Izvest. Akad. Nauk SSSR, Khim. Ser.* 1970, 8.

(The compilation based on this reference is included among the aqueous systems).

COMPONENTS:

(1) Potassium tetraphenylborate (1-); $KC_{24}H_{20}B$; [3244-41-5]

(2) 2-Propanone (acetone); C_3H_6O; [67-64-1]

(3) Water; H_2O; [7732-18-5]

ORIGINAL MEASUREMENTS:

Scott, A. D.; Hunziker, H. H.; Reed, M. G. *Chemist-Analyst* 1959, *48*, 11-12.

VARIABLES:

Acetone-water composition.
One temperature: 28°C

PREPARED BY:

Orest Popovych

EXPERIMENTAL VALUES:

The solubility of $KBPh_4$ in acetone-water mixtures at 28°C was reported in mg cm^{-3} by the authors and recalculated to mol dm^{-3} units by the compiler.

Vol % Acetone in Mixture*	Solubility of $KBPh_4$ mg cm^{-3}	mol dm^{-3}
16	0.4	1.1×10^{-3}
25	0.6	1.7×10^{-3}
32	1.2	3.4×10^{-3}
33	1.1	3.1×10^{-3}
38	2.9	8.1×10^{-3}
40	2.5	7.0×10^{-3}
44	4.9	1.4×10^{-2}
50	10.5, 8.9	2.93×10^{-2}, 2.49×10^{-2}
60	18.4	5.14×10^{-2}
70	27.0	7.54×10^{-2}
80	40.6	1.13×10^{-1}
90	53.0	1.48×10^{-1}
95	59.8	1.67×10^{-1}
99	49.3	1.38×10^{-1}
100	41.8	1.17×10^{-1}

*Percentage by volume based on combined volumes of the solvents.

AUXILIARY INFORMATION

METHOD/APPARATUS/PROCEDURE:

$KBPh_4$ was shaken for 16 hours in 20-ml portions of acetone-water mixtures or finely-ground samples shaken for 48 hours. Solutions with 50% acetone or more were analyzed by evaporation and weighing. In other solutions, the $KBPh_4$ was destroyed with aqua regia and the potassium determined by flame photometry.

SOURCE AND PURITY OF MATERIALS:

Reagent-grade acetone was used undried, and its water content (0.5% max.) was neither determined, nor corrected for. $KBPh_4$ prepared by metathesis of KCl and $NaBPh_4$ in solution acidified with HCl (presumably aqueous), washed with water saturated with $KBPh_4$ and air dried.

ESTIMATED ERROR:

Nothing specified.

REFERENCES:

COMPONENTS:

(1) Potassium tetraphenylborate (1-); $KC_{24}H_{20}B$; [3244-41-5]

(2) 2-Propanone (acetone); C_3H_6O; [67-64-1]

(3) Water; H_2O; [7732-18-5]

ORIGINAL MEASUREMENTS:

Kirgintsev, A. N.; Kozitskii, V. P. *Izvest. Akad. Nauk SSSR, Khim. Ser.* 1968, 1170-2.

VARIABLES:

Acetone-water composition
One temperature: 25.00°C

PREPARED BY:

Orest Popovych

EXPERIMENTAL VALUES:

The authors reported mass % of $KBPh_4$ in the saturated solutions, defined as grams of the salt in 100 cm^3 of the solution. The solubilities have been recalculated to mol dm^{-3} by the compiler.

% Water in acetone* vol %	Solubility of $KBPh_4$ (Mass/vol.)%	mol dm^{-3}
0.007	6.12	0.171
2	6.72	0.188
4	7.04	0.196
8	7.02	0.196
12	6.60	0.184
15	6.17	0.172
20	5.38	0.150
25	4.55	0.127
30	3.81	0.106
37	2.60	7.26×10^{-2}
45	1.60	4.47×10^{-2}
52	0.81	2.26×10^{-2}
60	0.35	9.8×10^{-3}
70	0.11	3.1×10^{-3}
80	0.037	1.03×10^{-3}

*Determined by weighing. Solvent volume was taken as the sum of the volumes of acetone and water, neglecting the effect of mixing.

AUXILIARY INFORMATION

METHOD/APPARATUS/PROCEDURE:

Evaporation and weighing. Saturated solutions were prepared by shaking the suspensions in a constant-temperature bath for 6 hours. Aliquots were removed through cotton plugs, weighed and the solvent removed by evaporation first under an IR lamp and then by oven-drying to constant weight at 105°C. The solid phase contained no solvent when recrystallized from acetone or acetone-water mixtures.

SOURCE AND PURITY OF MATERIALS:

$NaBPh_4$ ("analytical grade" from the Apolda Co., GDR) was purified by recrystallization from acetone-toluene, followed by dissolution in water, extraction with ether, and removal of the latter in vacuo. The purity of the final $NaBPh_4$ was no less than 99.6%. $KBPh_4$ was prepared by metathesis of $NaBPh_4$ with KCl and purified by double recrystallization from 20% water 80% acetone (probably by vol.). The acetone was slowly evaporated and the precipitate kept for a long time under high vacuum. Acetone was treated with $KMnO_4$ followed by triple fractionation. Final water content was 0.007 vol. % by Karl Fisher titration.

ESTIMATED ERROR:

Precision ±0.5%
Temperature control: ±0.05°C

COMPONENTS:

(1) Potassium tetraphenylborate (1-); $KC_{24}H_{20}B$; [3244-41-5]

(2) Water; H_2O; [7732-18-5]

(3) Dimethylsulfoxide; C_2H_6OS; [67-68-5]

ORIGINAL MEASUREMENTS:

Kundu, K. K.; Das, A. K.
J. Solution Chem. 1979, 259-65.

VARIABLES:

Composition of solvent at 25.0°C

PREPARED BY:

Orest Popovych

EXPERIMENTAL VALUES:

The solubility of potassium tetraphenylborate ($KBPh_4$) and its solubility (ion-activity) product were reported for three dimethylsulfoxide (DMSO)-water mixtures:

Mass % DMSO	10^3C/mol dm^{-3}	pK°_{s0} (volume units)
20	0.69	6.35
40	2.18	5.37
60	20.8	3.49

The activity coefficients $y_\pm$ were calculated from the Debye-Hückel equation in the extended form:

$$-\log y_\pm = \tfrac{1}{2}AC^{\frac{1}{2}}\left[(1 + \mathring{a}_+BC^{\frac{1}{2}})^{-1} + (1 + \mathring{a}_-BC^{\frac{1}{2}})^{-1}\right] + \log\left[(d - 0.001CM + 0.002CM_s)/d_s\right]\ldots$$

Where $\underline{C}$ is the solubility in mol dm^{-3}, $\underline{A}$ and $\underline{B}$ are the Debye-Hückel constants, $1.824 \times 10^6(\varepsilon_sT)^{-3/2}$ and $50.29(\varepsilon_sT)^{-1/2}$, respectively, $\mathring{a}_+$ is the ion-size parameter for the K^+ ion, taken as 0.3 nm, and $\mathring{a}_-$ is the ion-size parameter for the BPh_4^- ion, taken as 0.5 nm. M is the formula weight of the salt, M_s is the mean molecular weight of the solvent, $\underline{d}$ is the density of the solution, assumed to be approximately equal to that of the pure solvent $\underline{d}_s$.

AUXILIARY INFORMATION

METHOD/APPARATUS/PROCEDURE:

Saturated solutions were prepared by shaking for 3-4 h followed by equilibration in a thermostat. Filtered, weighed aliquots, after proper dilution with water, were analyzed by uv-spectrophotometry using a Beckman DU 2400 spectrophotometer. The shaking followed by thermostatting was repeated at 3-4-day intervals until constant absorption was obtained, which required 2-4 weeks.

SOURCE AND PURITY OF MATERIALS:

DMSO was purified by a literature method (1). $KBPh_4$ was prepared and purified as described by Popovych and Friedman (2) (see compilation for $KBPh_4$ in water).

ESTIMATED ERROR:

Temperature ±0.1°C
Precision in solubility: ±2%

REFERENCES:

(1) Das, A. K.; Kundu, K. K. *J. Chem. Soc. Faraday Trans. 1* 1973, *69*, 730.

(2) Popovych, O.; Friedman, R. M. *J. Phys. Chem.* 1966, *70*, 1671.

COMPONENTS:

(1) Potassium tetraphenylborate (1-); $KC_{24}H_{20}B$; [3244-41-5]
(2) Lithium chloride; LiCl; [7447-41-8]
(3) Ethanol; C_2H_6O; [64-17-5]
(4) Water; H_2O; [7732-18-5]

ORIGINAL MEASUREMENTS:

Dill, A. J.; Popovych, O.
J. Chem. Eng. Data 1969, *14*, 240-3.

VARIABLES:

Ethanol-water composition. LiCl concentration varied from 0 to 200 times molar solubility of $KC_{24}H_{20}B$. One temperature: 25.00°C

PREPARED BY:

Orest Popovych

EXPERIMENTAL VALUES:

The authors report the solubility C of potassium tetraphenylborate ($KBPh_4$) in ethanol-water mixtures without added LiCl. The solubilities with added LiCl are not reported, only the activity coefficients calculated from the variation of the above solubility as a function of ionic strength varied by means of LiCl.

Mass% ethanol in water	Solubility of $KBPh_4$, 10^3C/mol dm^{-3}
100.0	0.504
90.0	1.09
80.0	2.11
78.1	2.35
70.0	2.89
60.0	2.80
50.0*	2.37
46.0	2.08
40.0	1.33
30.0	0.670
20.0	0.340
10.0	0.220

*Graphically interpolated

continued ...

AUXILIARY INFORMATION

METHOD/APPARATUS/PROCEDURE:

Ultraviolet spectrophotometry using a Cary Model 14 spectrophotometer. Saturated solutions were prepared by shaking suspensions of $KBPh_4$ in water-jacketed flasks. A solution was considered saturated when successive weekly analses agreed to about 1%. This required about 2 weeks of equilibration for solutions without added LiCl and one month for solutions with added LiCl. Saturated solutions were filtered and analyzed spectrophotometrically using absorption coefficients characteristic of each solvent. All work was carried out in deaerated containers and solvents.

SOURCE AND PURITY OF MATERIALS:

$KBPh_4$ was prepared from $NaBPh_4$ (Fisher, 99.7%) by metathesis with KCl; it was recrystallized three times from 3:1 acetone-water and dried in vacuo at 80°C. Baker analyzed LiCl was doubly recrystallized from conductivity water and dried for 48 hours at 110°C. It was stored and transferred in a dry box. USP 95% ethanol was doubly distilled. USP absolute ethanol was refluxed

ESTIMATED ERROR: (For the solubility)

Precision ±1%
Accuracy ±3% (authors)
Temperature: ±0.01°C.

COMMENTS:

The results reported in this study can be considered as tentative values.

REFERENCES:

COMPONENTS:

(1) Potassium tetraphenylborate (1-); $KC_{24}H_{20}B$; [3244-41-5]
(2) Lithium chloride; LiCl; [7447-41-8]
(3) Ethanol; C_2H_6O; [64-17-5]
(4) Water; H_2O; [7732-18-5]

ORIGINAL MEASUREMENTS:

Dill, A. J.; Popovych, O.
J. Chem. Eng. Data 1969, *14*, 240-3.

COMMENTS AND/OR ADDITIONAL DATA

EXPERIMENTAL VALUES:

The authors determined the mean molar activity coefficients of potassium tetraphenylborate ($KBPh_4$) in ethanol-water mixtures from the variation of solubility as a function of the ionic strength varied by means of LiCl. The following equations were used:

$$\log \frac{\alpha_I C_I}{\alpha_o C_o} = \log y_{\pm,o} - \log y_{\pm,I} \quad \text{and} \quad -\log y_{\pm,I} = A_1 I^{1/2} + A_2 I + A_3 I^{3/2} + \ldots$$

where α's are the degrees of ionic association, $y_{\pm}$ is the mean molar activity coefficient ($f_{\pm}$ in the original), and the subscripts o and I denote pure solvent and ionic strength I as determined by the sum of LiCl and $KBPh_4$. At any ionic strength I, the activity coefficient can be calculated from the known A-coefficients. Values of α were calculated from the association constant K_A (1) using the same equation as in the compilation for $KBPh_4$ in methanol (2). C_o is the solubility in the pure solvent.

Mass% ethanol in water	α_o	$y_{\pm,o}$	A_1	A_2	$K^\circ_{s0} = (C_o\alpha_o y_{\pm,o})^2$, $mol^2\ dm^{-6}$ (compiler)
100.0	0.949	0.849	3.90	-7.55	1.65×10^{-7}
78.1	0.986	0.819	2.25	-3.47	3.61×10^{-6}
60.6	0.998	-----	1.30	-3.11	7.71×10^{-6}
38.4	1.000	-----	0.749	-0.820	1.58×10^{-6}
30.0*	1.000	0.958	----	-----	4.12×10^{-7}
20.0*	1.000	0.972	----	-----	1.09×10^{-7}
10.0*	1.000	0.981	----	-----	4.66×10^{-8}

*Activity coefficients calculated from the Debye-Hückel limiting law.

AUXILIARY INFORMATION

METHOD/APPARATUS/PROCEDURE:

SOURCE AND PURITY OF MATERIALS:...continued

over magnesium ethoxide for 12 hours under nitrogen and then distilled, collecting the middle fraction. Deionized water with a specific conductance of $3 \times 10^{-7}\ ohm^{-1}cm^{-1}$ was used. The exact mass% composition of ethanol-water mixtures was determined from the densities of the mixture and literature data (2).

ESTIMATED ERROR:

REFERENCES:

(1) Dill, A. J.; Popovych, O. *J. Chem. Eng. Data* 1969, *14*, 156.
(2) Popovych, O.; Friedman, R. M. *J. Phys. Chem.* 1966, *70*, 1671.
(3) Osborne, N. S.; McKelvey, E. C.; Bearce, H. W. *J. Wash. Acad. Sci.* 1912, *2*, 95.

COMPONENTS:
(1) Potassium tetraphenylborate (1-); $KC_{24}H_{20}B$; [3244-41-5]
(2) Lithium chloride; LiCl; [7447-41-8]
(3) Sodium hydroxide; NaOH; [1310-73-2]
(4) Methanol; CH_4O; [67-56-1]
(5) Water; H_2O; [7732-18-5]

ORIGINAL MEASUREMENTS:
LaBrocca, P. J.; Phillips, R.; Goldberg, S. S.; Popovych, O.
J. Chem. Eng. Data 1979, *24*, 215-8. (including Supplementary Material).

VARIABLES: Methanol-water composition. LiCl concentration varied from 0 to 10^3 times the solubility of $KBPh_4$ in mol dm^{-3}. One temperature: 25.00°C.

PREPARED BY:
Orest Popovych

EXPERIMENTAL VALUES:

The solubility of $KBPh_4$ in the presence of 2×10^{-5} mol dm^{-3} NaOH was reported in the absence of LiCl in the following methanol-water mixtures:

Mass % methanol in water	Solubility of $KBPh_4$, 10^3C/mol dm^{-3}
89.4	2.22
79.7	1.92
69.6	1.64
58.8	1.20
50.8	1.04
40.0	0.586
29.8	0.372
20.0	0.261
9.8	0.220

AUXILIARY INFORMATION

METHOD/APPARATUS/PROCEDURE: Ultraviolet spectrophotometry using a Cary Model 17 spectrophotometer. Saturated solutions were prepared by shaking the suspensions in water-jacketed flasks. A solution was considered saturated when successive weekly analyses agreed to about 1%. Saturated solutions were filtered and analyzed spectrophotometrically using absorption coefficients characteristic of each solvent. All work was carried out in deaerated containers and solvents.

SOURCE AND PURITY OF MATERIALS: $KBPh_4$ was prepared from $NaBPh_4$ (Fisher, 99.7%) by metathesis with KCl; it was recrystallized three times from 3:1 acetone-water and dried in vacuo at 80°C. Baker analyzed LiCl was doubly recrystallized from conductivity water and dried for 48 hrs. at 110°C. It was stored and transferred in a dry box. Certified ACS spectranalyzed methanol (Fisher Scientific Co.) was used without further purification. The densities of methanol-water mixtures were determined gravimetrically and their mass % obtained from literature data (1).

COMMENTS:
The result in this study can be designated as tentative values.

ESTIMATED ERROR: Precision ±1% (in the solubility)
Temperature control: ±0.01°C

REFERENCES:
(1) Bates, R. G.; Robinson, R. A. in *Chemical Physics of Ionic Solutions* Conway, B. E.; Barradas, R. G., Eds. Wiley. New York. 1966. Chapter 12.

COMPONENTS:	ORIGINAL MEASUREMENTS:
(1) Potassium tetraphenylborate (1-); $KC_{24}H_{20}B$; [3244-41-5] (2) Lithium chloride; LiCl; [7447-41-8] (3) Sodium hydroxide; NaOH; [1310-73-2] (4) Methanol; CH_4O; [67-56-1] (5) Water; H_2O; [7732-18-5]	LaBrocca, P. J.; Phillips, R.; Goldberg, S. S.; Popovych, O. *J. Chem. Eng. Data* 1979, *24*, 215-8. (including Supplementary Material).

EXPERIMENTAL VALUES:

Supplementary Material

Solubility of Potassium Tetraphenylborate in Methanol-Water Mixtures, c_{BPh_4}, as a Function of LiCl Concentration (all concentrations in mol dm^{-3})

89.4 Mass % Methanol		79.7 Mass % Methanol		58.8 Mass % Methanol	
$10^3c_{BPh_4}$	10^1c_{LiCl}	$10^3c_{BPh_4}$	10^1c_{LiCl}	$10^3c_{BPh_4}$	10^1c_{LiCl}
2.072	0	1.919	0	1.191	0
2.828	0.2261	2.378	0.2049	1.349	0.1108
3.460	1.131	2.690	0.6147	1.458	0.3324
3.601	1.583	2.826	1.024	1.532	0.5540
3.734	2.261	2.927	1.434	1.565	0.7756
3.979	4.522	3.069	2.049	1.624	1.108
3.544	9.044	3.143	4.098	1.682	2.216
3.323	11.31	3.105	6.147	1.702	3.324
2.855	15.83	2.970	8.196	1.698	4.432
2.624	18.09	2.860	10.24	1.671	5.540
		2.594	14.34	1.646	7.756
				1.505	11.08

50.8 Mass % Methanol		40.0 Mass % Methanol		29.8 Mass % Methanol	
$10^3c_{BPh_4}$	10^2c_{LiCl}	$10^4c_{BPh_4}$	10^2c_{LiCl}	$10^4c_{BPh_4}$	10^2c_{LiCl}
1.090	0	5.693	0	3.714	0
1.130	0.9150	6.104	0.5902	4.025	1.059
1.250	1.830	6.328	1.771	4.146	1.765
1.313	2.745	6.538	2.951	4.224	2.471
1.429	3.660	6.686	4.131	4.428	7.060
1.424	4.575	6.881	5.902	4.545	10.59
1.584	27.45	7.134	11.80	4.608	14.12
1.546	45.75	7.272	17.71	4.584	17.65
		7.277	23.61		
		7.301	29.51		
		7.158	41.31		
		7.072	47.22		

20.0 Mass % Methanol		9.8 Mass % Methanol	
$10^4c_{BPh_4}$	10^2c_{LiCl}	$10^4c_{BPh_4}$	10^2c_{LiCl}
2.571	0	2.177	0
2.728	0.2817	2.344	1.032
2.846	0.8451	2.373	1.465
2.831	1.409	3.388	2.064
2.949	1.972	2.486	4.128
2.935	2.817	2.491	6.192
3.033	5.634	2.604	8.256
3.072	8.451	2.589	10.32
3.131	11.27	2.648	14.45
3.195	14.09		
3.171	22.54		

COMPONENTS:

(1) Potassium tetraphenylborate (1-); $KC_{24}H_{20}B$; [3244-41-5]
(2) Lithium chloride; LiCl; [7447-41-8]
(3) Sodium hydroxide; NaOH; [1310-73-2]
(4) Methanol; CH_4O; [67-56-1]
(5) Water; H_2O; [7732-18-5]

ORIGINAL MEASUREMENTS:

LaBrocca, P. J.; Phillips, R.; Goldberg, S. S.; Popovych, O.
J. Chem. Eng. Data 1979, *24*, 215-8.

(including Supplementary Material)

continuation ... COMMENTS AND/OR ADDITIONAL DATA

EXPERIMENTAL VALUES:

From the variation of the solubility of $KBPh_4$ as a function of ionic strength varied by means of LiCl, the authors determined the mean ionic activity coefficients of $KBPh_4$ in the methanol-water mixtures using the following equations:

$$\log C_I/C_o = \log y_{\pm,o} - \log y_{\pm,I} \quad \text{and} \quad -\log y_{\pm,I} = A_1 I^{1/2} + A_2 I + A_3 I^{3/2} + \ldots$$

where $y_{\pm}$ is the mean molar activity coefficient ($f_{\pm}$ in the original) and the subscripts $\underline{o}$ and $\underline{I}$ denote solutions without and with added LiCl. Complete dissociation was assumed for all methanol-water mixtures, since $KBPh_4$ is practically unassociated even in pure methanol (1). At any ionic strength I (in mol dm^{-3}), the solubility C_I and the activity coefficient can be calculated from the A-coefficients characteristic of the methanol-water mixture, which are tabulated:

Mass % methanol in water	$y_{\pm,o}$	A_1	A_2	A_3	A_4	$K^\circ_{s0}/mol^2\ dm^{-6}$*
89.4	0.882 ± 0.032	1.38	-2.04	1.28	-0.343	(3.84 ± 0.20) x 10^{-6}
79.7	0.854 ± 0.020	1.21	-1.74	0.819	----	(2.69 ± 0.10) x 10^{-6}
69.6	0.913 ± 0.004	1.06	-2.28	2.70	-1.40	(2.24 ± 0.03) x 10^{-6}
58.8	0.935 ± 0.029	0.905	-1.70	1.52	-0.574	(1.26 ± 0.06) x 10^{-6}
50.8	0.918 ± 0.038	0.847	-0.849	----	----	(9.12 ± 0.55) x 10^{-7}
40.0	0.964 ± 0.003	0.697	-1.72	2.50	-1.70	(3.19 ± 0.05) x 10^{-7}
29.8	0.976 ± 0.006	0.688	-1.69	1.94	----	(1.30 ± 0.03) x 10^{-7}
20.0	0.988 ± 0.014	0.558	-1.11	0.750	----	(6.65 ± 0.17) x 10^{-8}
9.8	0.981 ± 0.017	0.585	-2.60	7.95	-9.16	(4.66 ± 0.13) x 10^{-8}

$^*K^\circ_{s0} = (C_o y_{\pm,o})^2$.

AUXILIARY INFORMATION

METHOD/APPARATUS/PROCEDURE:

SOURCE AND PURITY OF MATERIALS:

ESTIMATED ERROR:

The absolute precision in the $y_{\pm,o}$ and K°_{s0} values is indicated above.

REFERENCES:

(1) Popovych, O.; Friedman, R.M. *J. Phys. Chem.* 1966, *70*, 1671-3.

COMPONENTS:

(1) Potassium tetraphenylborate (1-); $KC_{24}H_{20}B$; [3244-41-5]

(2) Water; H_2O; [7732-18-5]

(3) Urea; CH_4ON_2; [57-13-6]

ORIGINAL MEASUREMENTS:

Kundu, K. K.; Das, A. K.
J. Solution Chem. 1979, 259-65.

VARIABLES:

Composition of solvent at 25.0°C

PREPARED BY:

Orest Popovych

EXPERIMENTAL VALUES:

The solubility of potassium tetraphenylborate ($KBPh_4$) and its solubility (ion-activity) product were reported for three urea-water mixtures:

Mass % Urea	10^4C/mol dm^{-3}	pK°_{s0} (volume units)
11.52	1.3	7.77
20.31	1.5	7.65
29.64	1.8	7.50
36.83	1.9	7.45

The activity coefficients $y_{\pm}$ were calculated from an extended Debye-Hückel equation in the form:

$$-\log y_{\pm} = \tfrac{1}{2}AC^{\frac{1}{2}} [(1 + \mathring{a}_{+}BC^{\frac{1}{2}})^{-1} + (1 + \mathring{a}_{-}BC^{\frac{1}{2}})^{-1}] + \log [(d - 0.001CM + 0.002CM_s)/d_s] \cdots$$

where $\underline{C}$ is the solubility in mol dm^{-3}, $\underline{A}$ and $\underline{B}$ are the Debye-Hückel constants, $1.824 \times 10^6(\varepsilon_s T)^{-3/2}$ and $50.29(\varepsilon_s T)^{-1/2}$, respectively, $\mathring{a}_{+}$ is the ion-size parameter for the K^+ ion, taken as 0.3 nm, and $\mathring{a}_{-}$ is the ion-size parameter for the BPh_4^- ion, taken as 0.5 nm. M is the formula weight of the salt, M_s is the mean molecular weight of the solvent, $\underline{d}$ is the density of the solution, assumed to be approximately equal to that of the pure solvent, $\underline{d}_s$.

AUXILIARY INFORMATION

METHOD/APPARATUS/PROCEDURE:

Saturated solutions were prepared by shaking for 3-4 h followed by equilibration in a thermostat. Filtered, weighed aliquots, after proper dilution with water, were analyzed by uv-spectrophotometry using a Beckman DU 2400 spectrophotometer. The shaking followed by thermostatting was repeated at 3-4 day intervals until constant absorption was obtained, which required 2-4 weeks.

SOURCE AND PURITY OF MATERIALS:

Urea was purified by a literature method (1). $KBPh_4$ was prepared and purified as described by Popovych and Friedman (2) (see compilation for $KBPh_4$ in water).

ESTIMATED ERROR:

Temperature: ±0.1°C
Precision in solubility: ±2%

REFERENCES:

(1) Kundu, K. K.; Majumdar, K. *J. Chem. Soc. Faraday Trans. 1* 1973, *69*, 807.
(2) Popovych, O.; Friedman, R. M. *J. Phys. Chem.* 1966, *70*, 1671.

COMPONENTS:

(1) Potassium tetraphenylborate (1-); $KC_{24}H_{20}B$; [3244-41-5]

(2) Acetonitrile; C_2H_3N; [75-05-8]

EVALUATOR:

Orest Popovych, Department of Chemistry, City University of New York, Brooklyn College, Brooklyn, N. Y. 11210, U. S. A.
September 1979

CRITICAL EVALUATION:

The solubility of potassium tetraphenylborate ($KBPh_4$) in acetonitrile was reported as such by Kolthoff and Chantooni (1) as well as by Popovych et al. (2). The solubility products of $KBPh_4$ in acetonitrile were reported in the above two studies and also in the two articles by Alexander and Parker (3, 4). All determinations were at 298 K.

Excellent agreement exists between the solubility value obtained by Kolthoff and Chantooni (1) from evaporation and weighing, C = 5.40 x 10^{-2} mol dm^{-3} and the value C = 5.33 x 10^{-2} mol dm^{-3} determined by UV-spectrophotometry by Popovych et al. (2). The latter study was carried out under temperature control to 0.01°C in the constant-temperature bath from which water was circulated through jacketed flasks containing the suspensions. Saturation was ascertained by successive analyses days apart until the solubilities agreed to 1% or better. Unfortunately, no experimental details are available in the Kolthoff and Chantooni (1) article as far as the temperature control and saturation attainment is concerned. Nevertheless, if we accept the number of significant figures retained in the result, it is possible to average the values 5.33 x 10^{-2} mol dm^{-3} and 5.40 x 10^{-2} mol dm^{-3} to arrive at the recommended value for the solubility of $KBPh_4$ in acetonitrile at 298 K: (5.36 ± 0.05) x 10^{-2} mol dm^{-3}. The indicated precision is that governing the UV-analysis for the BPh_4^- concentration. There are no data on the precision of the analysis by the method of evaporation and weighing.

Alexander and Parker report the formal (concentration) solubility product of $KBPh_4$ in acetonitrile as pK_{s0} = 2.7 (3) and 2.4 (4) (in volume units) in the two successive studies. Because these are concentration products, the values of the corresponding solubilities can be calculated from them simply as $(K_{s0})^{1/2}$. When pK_{s0} = 2.7, the solubility is 4.5 x 10^{-2} mol dm^{-3}. The solubility corresponding to the pK_{s0} = 2.4 is 6.3 x 10^{-2} mol dm^{-3}. However, it should be noted that in the second study the authors estimate a precision of ±0.2 pK units, which means that the solubility derived from it could range from 5 x 10^{-2} to 8 x 10^{-2} mol dm^{-3}. Clearly, these results cannot be compared in precision with the recommended values stated above. While in the case of solubility of $KBPh_4$ in acetonitrile it is possible to recommend a value, this is unfortunately not so in the case of the thermodynamic solubility product. The latter was estimated both by Kolthoff and Chantooni (1) as well as by Popovych et al. (2) using calculated activity coefficients and at the concentration involved the differences in the activity corrections can be appreciable. By employing an unspecified form of the Guggenheim equation, Kolthoff and Chantooni calculated a pK°_{s0} = 3.2 (K°_{s0} units are mol^2 dm^{-6} in this evaluation). Popovych et al. (2) by suing a Debye-Hückel equation shown on the compilation sheet obtained for the mean ionic activity coefficient $y_\pm^2$ = 0.298, from which K°_{s0} = 8.47 x 10^{-4} mol^2 dm^{-6} and the pK°_{s0} = 3.07. The above activity coefficient was calculated using ion-size parameters å = 0.3 nm for the K^+ ion and å = 0.5 nm for the BPh_4^- ion. The latter, however, may be too small. For example, Kolthoff and Chantooni (5) used an ion-size parameter of 1.2 nm for the BPh_4^-. Applying this value, the mean ionic activity coefficient in acetonitrile becomes $y_\pm^2$ = 0.359, the K°_{s0} becomes 1.02 x 10^{-3} and the pK°_{s0} = 2.99. Thus, for approximate work, one can choose a pK°_{s0} value of about 3.1 ± 0.1, but it can be described as no better than tentative. A recommended value for solubility product of $KBPh_4$ in acetonitrile must await an experimental determination of the activity coefficients.

COMPONENTS:

(1) Potassium tetraphenylborate (1-); $KC_{24}H_{20}B$; [3244-41-5]

(2) Acetonitrile; C_2H_3N; [75-05-8]

EVALUATOR:

Orest Popovych, Department of Chemistry, City University of New York, Brooklyn College, Brooklyn, N. Y. 11210, U. S. A.
September 1979

CRITICAL EVALUATION: (continued)

REFERENCES:

1. Kolthoff, I. M.; Chantooni, M. K., Jr. *J. Phys. Chem.* 1972, *76*, 2024.
2. Popovych, O.; Gibofsky, A.; Berne, D. H. *Anal. Chem.* 1972, *44*, 811.
3. Alexander, R.; Parker, A. J. *J. Am. Chem. Soc.* 1967, *89*, 5549.
4. Parker, A. J.; Alexander, R. *J. Am. Chem. Soc.* 1968, *90*, 3313.
5. Kolthoff, I. M.; Chantooni, M. K., Jr. *Anal. Chem.* 1972, *44*, 194.

COMPONENTS:

(1) Potassium tetraphenylborate (1-); $KC_{24}H_{20}B$; [3244-41-5]

(2) Acetonitrile; C_2H_3N; [75-05-8]

ORIGINAL MEASUREMENTS:

Kolthoff, I. M.; Chantooni, M. K., Jr. *J. Phys. Chem.* 1972, *76*, 2024-34.

VARIABLES:

One temperature: 25°C

PREPARED BY:

Orest Popovych

EXPERIMENTAL VALUES:

The authors report the solubility of potassium tetraphenylborate ($KBPh_4$) in acetonitrile as:

$$C = 5.40 \times 10^{-2} \text{ mol dm}^{-3}.$$

Assuming complete dissociation, and calculating the mean ionic activity coefficient from the Guggenheim equation*, the authors report as the solubility product of $KBPh_4$:

$$pK^\circ_{s0} = 3.2 \ (K^\circ_{s0} \text{ units are mol}^2 \text{ dm}^{-6}).$$

*Not shown.

AUXILIARY INFORMATION

METHOD/APPARATUS/PROCEDURE:

Evaporation of a saturated solution and weighing. No other details.

SOURCE AND PURITY OF MATERIALS:

Acetonitrile was purified very thoroughly by a literature method (1). Sodium tetraphenylborate (Aldrich puriss. grade) was purified according to the method of Popov and Humphrey (2). $KBPh_4$ was prepared by metathesis of KCl with $NaBPh_4$, referring to the procedure described in the compilation for $KBPh_4$ in methanol.

ESTIMATED ERROR:

Nothing specified.

REFERENCES:

(1) Kolthoff, I. M.; Bruckenstein, S. Chantooni, M. K., Jr. *J. Am. Chem. Soc.* 1961, *83*, 3927.
(2) Popov, A. I.; Humphrey, R. *J. Am. Chem. Soc.* 1959, *81*, 2043.

COMPONENTS:

(1) Potassium tetraphenylborate (1-); $KC_{24}H_{20}B$; [3244-41-5]

(2) Acetonitrile; C_2H_3N; [75-05-8]

ORIGINAL MEASUREMENTS:

Popovych, O.; Gibofsky, A.; Berne, D. H. *Anal. Chem.* 1972, *44*, 811-17.

VARIABLES:

One temperature: 25.00°C

PREPARED BY:

Orest Popovych

EXPERIMENTAL VALUES:

The solubility was reported as $C_{BPh_4} = 5.33 \times 10^{-2}$ mol dm^{-3}.

The mean molar ionic activity coefficient was calculated using the relationship:

$$-\log y_{\pm}^2 = \frac{1.64C^{1/2}}{1 + 0.485\mathring{a}C^{1/2}}$$

Adopting $\mathring{a}$ = 0.5 nm for BPh_4^- and $\mathring{a}$ = 0.3 nm for K^+, the value of $y_{\pm}^2$ = 0.298 and the pK_{s0}° derived from it is 2.85 (molal scale), i. e., K_{s0}° units are mol^2 kg^{-2}. pK_{s0}° value on the molar scale (K_{s0}° units of mol^2 dm^{-6}) was not reported, but can be calculated from the molal value via the solvent density, which was 0.777 g ml^{-1}. pK_{s0}° (molar scale) = 3.07. Complete dissociation was assumed, which is generally true for most electrolytes in acetonitrile. Also reported were the molar absorption coefficients for the BPh_4^- ion: 3203 and 2082 dm^3 (cm mol)$^{-1}$ at 266 nm and 274 nm, respectively.

AUXILIARY INFORMATION

METHOD/APPARATUS/PROCEDURE:

Ultraviolet spectrophotometry using a Cary Model 14 spectrophotometer. Saturation achieved by shaking the salt suspensions for 2 weeks in water-jacketed flasks. Solutions filtered and analyzed at 266 and 274 nm using the absorption coefficients specified above. All solutions and containers were deaerated. Differential thermal analysis showed absence of crystal solvates.

SOURCE AND PURITY OF MATERIALS:

Acetonitrile (Matheson, spectroquality) was refluxed for 24 hrs over CaH_2 and fractionally distilled. $KBPh_4$ was prepared from $NaBPh_4$ (Fisher, 99.7%) and KCl by metathesis in aqueous solution. It was recrystallized three times from 3:1 acetone-water and dried in vacuo at 80°C.

ESTIMATED ERROR:

Precision ±1% (rel.)
Accuracy ±3% (rel.)
Temperature control: ±0.01°C

REFERENCES:

COMPONENTS:

(1) Potassium tetraphenylborate (1-); $KC_{24}H_{20}B$; [3244-41-5]

(2) Acetonitrile; C_2H_3N; [75-05-8]

ORIGINAL MEASUREMENTS:

Alexander, R.; Parker, A. J.
J. Am. Chem. Soc. 1967, *89*, 5549-51.

VARIABLES:

One temperature: 25°C

PREPARED BY:

Orest Popovych

EXPERIMENTAL VALUES:

The solubility product was calculated using concentrations. The authors reported pK_{s0} = 2.7, where the solubility product is in units of $mol^2\ dm^{-6}$.

AUXILIARY INFORMATION

METHOD/APPARATUS/PROCEDURE:

Spectrophotometry of the solutions saturated under nitrogen or potentiometric titration of the anion with $AgNO_3$. No details.

SOURCE AND PURITY OF MATERIALS:

Not stated.

ESTIMATED ERROR:

Nothing specified.

REFERENCES:

COMPONENTS:

(1) Potassium tetraphenylborate (1-); $KC_{24}H_{20}B$; [3244-41-5]

(2) Acetonitrile; C_2H_3N; [75-05-8]

ORIGINAL MEASUREMENTS:

Parker, A. J.; Alexander, R.
J. Am. Chem. Soc. 1968, *90*, 3313-9.

VARIABLES:

One temperature: 25°C

PREPARED BY:

Orest Popovych

EXPERIMENTAL VALUES:

The formal (concentration) solubility product of $KBPh_4$ in acetonitrile was reported as:

$$pK_{s0} = 2.4 \ (K_{s0} \text{ units are mol}^2 \text{ dm}^{-6}).$$

AUXILIARY INFORMATION

METHOD/APPARATUS/PROCEDURE:

UV spectrophotometry on solutions saturated under nitrogen, using a Unicam SP500 spectrophotometer. Saturated solutions were prepared by shaking for 24 hours at 35°C and then for a further 24 hours at 25°C.

SOURCE AND PURITY OF MATERIALS:

The purification of materials has been described in the literature (1-3).

ESTIMATED ERROR:

Absolute precision was estimated to be ±0.2 pK units.
Temperature control unspecified.

REFERENCES:

(1) Clare, B. W.; Cook, D.; Ko, E. C. F.;Mac, Y. C.; Parker, A. J. *J. Am. Chem. Soc.* 1966, *88*, 1911.
(2) Alexander, R.; Ko, E. C. F.; Mac, Y. C.; Parker, A. J. *J. Am. Chem. Soc.* 1967, *89*, 3703.
(3) Parker, A. J. *J. Chem. Soc. A* 1966, 220.

COMPONENTS:

(1) Potassium tetraphenylborate (1-); $KC_{24}H_{20}B$; [3244-41-5]

(2) Formamide; CH_3NO; [75-12-7]

ORIGINAL MEASUREMENTS:

Parker, A. J.; Alexander, R.
J. Am. Chem. Soc. 1968, *90*, 3313-9.

VARIABLES:

One temperature: 25°C

PREPARED BY:

Orest Popovych

EXPERIMENTAL VALUES:

The formal (concentration) solubility product of $KBPh_4$ in formamide was reported as:

$$pK_{s0} = 2.8 \ (K_{s0} \text{ units are } mol^2 \ dm^{-6}).$$

AUXILIARY INFORMATION

METHOD/APPARATUS/PROCEDURE:

UV spectrophotometry on solutions saturated under nitrogen, using a Unicam SP500 spectrophotometer. Saturated solutions were prepared by shaking for 24 hours at 35°C and then for a further 24 hours at 25°C.

SOURCE AND PURITY OF MATERIALS:

The purification of the materials has been described in the literature (1-3).

ESTIMATED ERROR:

Absolute precision was estimated to be ±0.2 pK units.

REFERENCES:

(1) Clare, B. W.;Cook, D.; Ko, E. C. F.; Mac, Y. C.; Parker, A. J *J. Am. Chem. Soc.* 1966, *88*, 1911.
(2) Alexander, R.; Ko, E. C. F.; Mac, Y. C.; Parker, A. J. *J. Am. Chem. Soc.* 1967, *89*, 3703.
(3) Parker, A. J. *J. Chem. Soc. A* 1966, 220.

COMPONENTS:

(1) Potassium tetraphenylborate (1-); $KC_{24}H_{20}B$; [3244-41-5]

(2) Methanol; CH_4O; [67-56-1]

ORIGINAL MEASUREMENTS:

Popovych, O.; Friedman, R. M. *J. Phys. Chem.* 1966, *70*, 1671-3.

VARIABLES:

One temperature: 25.00°C

PREPARED BY:

Orest Popovych

EXPERIMENTAL VALUES:

The solubility of $KBPh_4$ in methanol was reported to be:

$$C = 3.11 \times 10^{-3} \text{ mol dm}^{-3}.$$

The solubility product, K°_{s0}, was calculated by the authors as $(C\alpha y_\pm)^2$, where α is the degree of dissociation and $y_\pm$, the mean ionic activity coefficient on the molar scale. α was calculated from a literature calue of the ion-pair association constant K_A = 22 $mol^{-1}dm^3$ (1), using the relationship:

$$\alpha = \frac{-1 + (1 + 4K_ACy_\pm{}^2)^{1/2}}{2K_ACy_\pm{}^2}$$

. The activity coefficient was estimated from the Debye-Hückel equation in the form:

$$-\log y_\pm{}^2 = \frac{3.803\ (C\alpha)^{1/2}}{1 + 0.5099\ \mathring{a}(C\alpha)^{1/2}}$$

using 0.55 nm as the value for the ion-size parameter $\mathring{a}$. The above calculations yielded α = 0.958 and $y_\pm{}^2$ = 0.660, from which the reported K°_{s0} = 5.86 x 10^{-6} mol^2 dm^{-6}.

Also reported were the molar absorption coefficients of the tetraphenylborate ion in methanol: 3.00 x 10^3 $dm^3(cm\ mol)^{-1}$ and 2.12 x 10^3 $dm^3(cm\ mol)^{-1}$ at 266 and 274 nm, respectively.

AUXILIARY INFORMATION

METHOD/APPARATUS/PROCEDURE:

Ultraviolet spectrophotometry using a Cary Model 14 spectrophotometer. Saturation achieved by shaking the salt suspensions for 2 weeks in water-jacketed flasks. Solutions filtered and analyzed at 266 and 274 nm using the above absorption coefficients. All solutions and containers were deaerated.

SOURCE AND PURITY OF MATERIALS:

$KBPh_4$ was prepared from $NaBPh_4$ (Fisher, 99.7%) and KCl by metathesis in aqueous solution. It was recrystallized three times from 3:1 acetone-water and dried in vacuo at 80°C. Methanol (Matheson, spectro grade) was refluxed over Al amalgam and distilled, rejecting the initial and final 10%.

COMMENTS:

The solubility and the K°_{s0} in this study can be designated as tentative values. However, the 3 significant digits in K°_{s0} are not justified in view of the uncertainty in the value of K_A, which was reported (1) as ranging from 6 to 35, with an average of 22. Taking this into account, the K°_{s0} should be expressed as (5.9 ± 0.3) x 10^{-6} mol^2 dm^{-6}.

ESTIMATED ERROR:

Not stated by the authors, but rel. precision in the solubility is known to be about ±1%.
Temperature control: ±0.01°C.

REFERENCES:

(1) Kunze, R. W.; Fuoss, R. M. *J. Phys. Chem.* 1963, *67*, 911.

COMPONENTS:

(1) Potassium tetraphenylborate (1-); $KC_{24}H_{20}B$; [3244-41-5]

(2) N-Methyl-2-pyrrolidinone (N-Methyl-2-pyrrolidone); C_5H_9NO; [872-50-4]

ORIGINAL MEASUREMENTS:

Virtanen, P. O. I.; Kerkelä, R. *Suomen Kemistilehti* 1969, *B42*, 29-33.

VARIABLES:

Two temperatures: 25.00°C and 45.00°C

PREPARED BY:

Orest Popovych

EXPERIMENTAL VALUES:

The solubility of $KBPh_4$ in N-methyl-2-pyrrolidone was reported to be 1.01 mol dm^{-3} at 25° C and 1.03 mol dm^{-3} at 45° C.

The corresponding solubility product at 25° C, calculated as the square of the solubility, was reported in the form pK_{s0} = -0.01, where K_{s0} units are mol^2 dm^{-6}. The pK_{s0} value at 45° C was not reported.

AUXILIARY INFORMATION

METHOD/APPARATUS/PROCEDURE:

The suspensions were shaken in thermostatted water-jacketed flasks for 1 day at 50°C, followed by 1 day at 25°C or 45°C, respectively. Saturated solutions were anayzed by precipitating the $KBPh_4$ from aliquots in aqueous solution.

SOURCE AND PURITY OF MATERIALS:

N-Methyl-2-pyrrolidone (General Aniline & Film Co.) was purified as in the literature (1). $KBPh_4$ was prepared by metathesis of KCl and $NaBPh_4$ in water, followed by double recrystallization from an acetone-water mixture and drying in vacuo.

ESTIMATED ERROR:

Not specified.
Temperature control: ±0.02°C

REFERENCES:

(1) Virtanen, P. O. I. *Suomen Kemistilehti* 1966, *B39*, 257.

COMPONENTS:

(1) Potassium tetraphenylborate (1-); $KC_{24}H_{20}B$; [3244-41-5]

(2) 2-Propanone (acetone); C_3H_6O; [67-64-1]

EVALUATOR:

Orest Popovych, Department of Chemistry, City University of New York, Brooklyn College, Brooklyn, N. Y. 11210, U. S. A.

September 1979

CRITICAL EVALUATION:

The solubility of potassium tetraphenylborate ($KBPh_4$) in acetone has been reported in three publications (1-3). Scott et al. (1) determined the solubility in acetone as part of their study of the solubility in acetone-water mixtures at 301 K (see compilation for acetone-water mixtures). Similarly, Kirgintsev and Kozitskii (2) included the acetone datum in their report on the solubilities of $KBPh_4$ in acetone-water mixtures at 298.15 K (see compilation for acetone-water mixtures). Subsequently, Kozitskii (3) published the solubilities of $KBPh_4$ in acetone at 273.15 K and 323.15 K. Thus, no comparison is available between data from two laboratories at any one temperature. However, solubilities at three different temperatures are available from the same laboratory (2, 3). All the available data are summarized in the Table below.

Tentative Values

Solubilities of $KBPh_4$ in Acetone at Different Temperatures

T/K	Solubility/mol dm^{-3}
273.15	0.196 (3)
298.15	0.171 (2)
301	0.117 (1)*
323.15	0.144 (3)

*Doubtful value

All of the above solubilities were determined by the method of evaporation and weighing, but the purity of the acetone employed and the temperature control were not the same in the different studies. The solubilities at the temperatures other than 301 K came from the same laboratory and were measured in thoroughly dried acetone (0.007 vol % of water), observing a temperature control of ±0.05°C (2, 3). On the other hand, the datum of Scott et al. (1) at 301 K was determined in acetone which may have contained up to 0.5% water (by volume?) and the solubility value of 0.117 mol dm^{-3} obtained by them clearly does not belong to the same population as the remaining three data points in the Table.

Since the activity correction would be too uncertain at the solubilities involved here, the corresponding solubility products were not estimated. Instead, a smoothing equation was obtained for the logarithm of the solubility S as a function of reciprocal absolute temperature, using the three data points at 273.15 K, 298.15 K and 323.15 K (all from the same laboratory (2, 3)):

$\log S = 235/(T/K) - 1.56$, with a $\sigma_y = 0.0062$ and a correlation coefficient of 0.994. The solubility value calculated from the above equation for 301 K is 0.165 mol dm^{-3}. This differs considerably from the 0.117 mol dm^{-3} value reported by Scott et al. (1). Since low concentrations of water in acetone lead to an <u>increase</u> in the solubility (2) (see compilation) for $KBPh_4$ in acetone-water mixtures based on Reference (2)), the low solubility value in the study by Scott et al. (1) cannot be rationalized on the basis of the wetness of their acetone. Of course, the unspecified degree of the temperature control and saturation control in the last study, as well as differences between the amounts of residue obtained from the solvent in different studies could account for the discrepancy between the results from the laboratory of Kirgintsev and Kozitskii (2,3) on the one hand and those of Scott et al. (1) on the other hand.

COMPONENTS:	EVALUATOR:
(1) Potassium tetraphenylborate (1-) $KC_{24}H_{20}B$; [3244-41-5] (2) 2-Propanone (acetone); C_3H_6O; [67-64-1]	Orest Popovych, Department of Chemistry, City University of New York, Brooklyn College, Brooklyn, N. Y. 11210, U. S. A. September 1979

CRITICAL EVALUATION: (continued)

In conclusion, the only thing that prevents this evaluator from designating the solubility data of Kirgintsev and Kositskii (2) and of Kositskii (3) as the recommended values is the fact that equilibrium for only 6 hours may have been insufficient for complete saturation. Thus, the solubility values at all temperatures except 301 K listed in the Table on the preceeding page should be considered as tentative values at this time.

REFERENCES:

1. Scott, A. D.; Hunziker, H. H.; Reed, M. G. *Chemist-Analyst* 1959, *48*, 11.
2. Kirgintsev, A. N.; Kozitskii, V. P. *Izvest. Akad. Nauk SSSR, Khim. Ser.* 1968, 1170.
3. Kozitskii, V. P. *Izvest. Akad. Nauk. SSSR, Khim. Ser.* 1970, 8.

COMPONENTS:

(1) Potassium tetraphenylborate (1-); $KC_{24}H_{20}B$; [3244-41-5]

(2) 2-Propanone (acetone); C_3H_6O; [67-64-1]

ORIGINAL MEASUREMENTS:

Kozitskii, V. P. *Izvest. Akad. Nauk SSSR, Khim. Ser.* 1970, 8-11.

VARIABLES:

Two temperatures: 0.00°C and 50.00°C

PREPARED BY:

Orest Popovych

EXPERIMENTAL VALUES:

The author reported the solubility of $KBPh_4$ in acetone* in units of mass %, defined as grams of the salt in 100 cm^3 of the saturated solution. The solubility was converted to units of mol dm^{-3} by the compiler.

Solubility of $KBPh_4$ in Acetone

0°C	7.04 mass % or 1.96×10^{-1} mol dm^{-3}
50°C	5.15 mass % or 1.44×10^{-1} mol dm^{-3}

*Containing 0.007% water by volume.

AUXILIARY INFORMATION

METHOD/APPARATUS/PROCEDURE:

Evaporation and weighing. Saturated solutions were prepared by stirring mechanically the suspensions in the thermostatted baths for the length of time indicated in the above Table. One liter of the filtrate from the saturated solution was dried by vacuum distillation (∿5 hrs keeping the temperature no higher than 40°C). The residue and the column were rinsed out many times with small portions of acetone, which was then collected, evaporated, and the residue, weighed.

SOURCE AND PURITY OF MATERIALS:

Absolute acetone (0.0065 vol % H_2O) was prepd by the same method as described on the compilation sheet for $KBPh_4$ in acetone-water at 25°C. $KBPh_4$ obtained by metathesis of KCl and $NaBPh_4$ was purified by double recrystallization from 3:1 acetone-water, evaporation of the acetone, washing of the crystals with water and ether and vacuum drying at 60°C. Water was doubly distilled.

ESTIMATED ERROR:

Not stated.
Temperature control: ±0.05°C

REFERENCES:

1. Kirgintsev, A. N.; Kozitskii, V. P. *Izvest. Akad. Nauk SSSR, Khim. Ser.* 1968, 1170.

COMPONENTS:

(1) Rubidium tetraphenylborate (1-); $RbC_{24}H_{20}B$; [5971-93-7]

(2) Water; H_2O; [7732-18-5]

EVALUATOR:

Orest Popovych, Department of Chemistry, City University of New York, Brooklyn College, Brooklyn, N. Y. 11210, U. S. A.
September 1979

CRITICAL EVALUATION:

Five publications containing original data on the solubility of rubidium tetraphenylborate ($RbBPh_4$) in aqueous solutions have been reviewed (1-5). Historically the first datam, the conductometrically determined solubility product of $RbBPh_4$, $K_{s0} = 8 \times 10^{-10}$ (presumably $mol^2\ dm^{-6}$) at 290 K, was rejected (no compilation sheet provided), because nothing was specified in that communication (1). Next was the radiometric determination by Geilman and Gebauhr (2), in which the solubility was reported as 4.4×10^{-5} mol dm^{-3} at 293.2 K. (The last value was recalculated from raw data in the article by the compiler). In pure water at 298 K, there are the data of Pflaum and Howick (3), where the solubility is given as 2.33×10^{-5} mol dm^{-3} and of Popovych et al. (4), where it is reported as $5.4_2 \times 10^{-5}$ mol dm^{-3}. The last two studies were carried out by UV-spectrophotometry, but their results are unfortunately in poor agreement. Because care was taken in the latter study to control the temperature in the bath to 0.01°C and to ensure saturation by successive analyses days apart until the results checked to 1% or better, the solubility value of 5.4×10^{-5} mol dm^{-3} is the most reliable we have at the moment at 298.15 K and it should be regarded as the tentative value. Pflaum and Howick (3), on the other hand, gave no details on the temperature control or the saturation procedure. The only other solubility datum at 298 K was determined in a buffer solution, not in pure water (5).

REFERENCES:

1. Rüdorff, W.; Zannier, H. *Angew. Chem.* 1952, *64*, 613.
2. Geilman, W.; Gebauhr, W. Z. *anal. Chem.* 1953, *139*, 161.
3. Pflaum, R. T.; Howick, L. C. *Anal. Chem.* 1956, *44*, 1542.
4. Popovych, O.; Gibofsky, A.; Berne, D. H. *Anal. Chem.* 1972, *44*, 811.
5. McClure, J. E.; Rechnitz, G. A. *Anal. Chem.* 1966, *38*, 136.

COMPONENTS:

(1) Rubidium tetraphenylborate (1-); $RbC_{24}H_{20}B$; [5971-93-7]

(2) Water; H_2O; [7732-18-5]

ORIGINAL MEASUREMENTS:

Geilman, W.; Gebauhr, W.
Z. anal. Chem. 1953, *139*, 161-81.

VARIABLES:

One temperature: 20°C

PREPARED BY:

Orest Popovych

EXPERIMENTAL VALUES:

The solubility is reported both as C_{Rb} = 0.380 mg cm^{-3} and as 4.5 x 10^{-5} mol dm^{-3}. However, inspection of the raw data (below) suggests that the authors consider only two figures to be significant, i. e. 0.38 mg cm^{-3} should be the solubility. Using 85.48 for the atomic mass of Rb, this compiler obtains for the solubility C_{Rb} = 4.4 x 10^{-5} mol dm^{-3}. Correspondingly, the solubility product, which the authors report simply as C_{Rb}^2 = 2.0 x 10^{-9} mol^2 dm^{-6}, should be K_{s0} = 1.9 x 10^{-9} mol^2 dm^{-6} (compiler). Also reported is the rate of dissolution of $RbBPh_4$ in water:

Time, hours	μg Rb/10 cm^3 of water
0.5	31.5
1.0	33.6
3.0	34.2
8.0	35.5
20.0	37.9
33.0	37.8

AUXILIARY INFORMATION

METHOD/APPARATUS/PROCEDURE:

Radiometric, using liquid-scintillation counting of ^{86}Rb. The radioactive rubidium obtained as the carbonate from the Harwell nuclear reactor was purified by precipitation with $HClO_4$ in the presence of 2 mg of Na_2HPO_4 followed by recrystallization The $RbClO_4$ solution was reacted with $NaBPh_4$, the resulting $RbBPh_4$ precipitate washed with water, mechanically shaken in water at 20°C and the filtrate analyzed radiometrically to constant activity. Apparatus was not specified.

SOURCE AND PURITY OF MATERIALS:

Not stated.

ESTIMATED ERROR:

Not specified. However, given the temperature control to ±0.5°C, the relative precision cannot be better than ±1-2% (compiler).

REFERENCES:

COMPONENTS:

(1) Rubidium tetraphenylborate (1-); $RbC_{24}H_{20}B$; [5971-93-7]

(2) Water; H_2O; [7732-18-5]

ORIGINAL MEASUREMENTS:

Pflaum, R. T.; Howick, L. C.
Anal. Chem. 1956, *28*, 1542-44.

VARIABLES:

One temperature: 25°C

PREPARED BY:

Orest Popovych

EXPERIMENTAL VALUES:

The solubility of $RbBPh_4$ in water was reported as 2.33×10^{-5} mol dm^{-3}.

AUXILIARY INFORMATION

METHOD/APPARATUS/PROCEDURE:

Ultraviolet spectrophotometry on a Cary Model 11 recording spectrophotometer. Saturated solutions were prepared in conductivity water by an unspecified procedure. Method of controlling the temperature was not stated. The concentration of BPh_4^- in saturated solutions was obtained from spectrophotometric measurements at 266 and 274 nm by applying the molar absorption coefficients of 3225 and 2100 dm^3 $(cm\ mol)^{-1}$, respectively. However, the above absorption coefficients were determined on acetonitrile solutions.

SOURCE AND PURITY OF MATERIALS:

$NaBPh_4$ (J. T. Baker Chemical Co.) was used as received for pptns, but was recrystallized from acetone-hexane mixt for detn of absorption coefficients. $RbBPh_4$ was prepd by metathesis of RbCl and $NaBPh_4$ and purified by recrystallization from a CH_3CN-H_2O mixt. CH_3CN (Matheson, Coleman & Bell) was treated with cold satd KOH, dried over anhydrous K_2CO_3 for 24 hrs., refluxed over P_2O_5 for 2-3 hrs. and then distilled from P_2O_5 in an all-glass apparatus. The fraction boiling at 81-81.5°C was retained. All other chemicals were of reagent grade quality.

ESTIMATED ERROR:

Nothing was specified, but the precision is likely to be ±1% (compiler).

COMPONENTS:

(1) Rubidium tetraphenylborate (1-); $RbC_{24}H_{20}B$; [5971-93-7]
(2) Water; H_2O; [7732-18-5]

ORIGINAL MEASUREMENTS:

Popovych, O.; Gibofsky, A.; Berne, D. H. *Anal. Chem.* 1972, *44*, 811-7.

VARIABLES:

One temperature: 25.00°C

PREPARED BY:

Orest Popovych

EXPERIMENTAL VALUES:

The solubility of $RbBPh_4$ was reported as:

$$c_{BPh_4} = 5.4_2 \times 10^{-5} \text{ mol dm}^{-3}.$$

Combining the above value with the mean molar activity coefficient calculated from the Debye-Hückel limiting law $-\log y_{\pm} = 0.509C^{\frac{1}{2}}$, the authors reported as the solubility product: $pK^{\circ}_{s0} = 8.54$ (K°_{s0} units are $mol^2\ kg^{-2}$).

AUXILIARY INFORMATION

METHOD/APPARATUS/PROCEDURE:

UV spectrophotometry using a Cary Model 14 spectrophotometer. Saturated solutions were prepared by shaking suspensions of $RbBPh_4$ in water-jacketed flasks. After about two weeks of shaking , the suspensions were filtered and the filtrates analyzed spectrophotometrically. The molar absorption coefficients of 3.25×10^3 and 2.06×10^3 $dm^3(cm\ mol)^{-1}$ at 266 nm and 274 nm, respectively, were used to compute the concentration of tetraphenylborate. All work was carried out in deaerated containers and solvents. Differential thermal analysis showed absence of crystal solvates.

SOURCE AND PURITY OF MATERIALS:

$RbBPh_4$ was prepared from $NaBPh_4$ (Fisher, 99.7%) by metathesis with RbCl; it was recrystallized three times from 3:1 acetone-water and dried in vacuo at 80°C.
Deionized water was redistilled.

ESTIMATED ERROR:

Precision ±2% (rel.)
Temperature control: ±0.01°C

REFERENCES:

COMPONENTS:

(1) Rubidium tetraphenylborate (1-); $RbC_{24}H_{20}B$; [5971-93-7]

(2) Tris(hydroxymethyl)aminomethane; $C_4H_{11}NO_3$; [77-86-1]

(3) Ethanoic acid (acetic acid); $C_2H_4O_2$; [64-19-7]

(4) Water; H_2O; [7732-18-5]

ORIGINAL MEASUREMENTS:

McClure, J. E.; Rechnitz, G. A. *Anal. Chem.* 1966, *38*, 136-9.

VARIABLES:

One temperature: 24.8°C

PREPARED BY:

Orest Popovych

EXPERIMENTAL VALUES:

The solubility of rubidium tetraphenylborate ($RbBPh_4$) in aqueous tris(hydroxymethyl)aminomethane (THAM) buffer at pH 5.1 was reported as:

$$6.7 \times 10^{-5} \text{ mol dm}^{-3}.$$

AUXILIARY INFORMATION

METHOD/APPARATUS/PROCEDURE:

UV-spectrophotometry using a Cary Model 11 spectrophotometer according to the procedure of Howick and Pflaum (1). No other details. In the cited procedure, saturated solutions were prepared both by agitating the suspensions at 25°C continuously and by agitating them first for a 0.5 hr at 40-50°C and then cooling to 25°C. The equilibrated solutions were filtered prior to analysis.

SOURCE AND PURITY OF MATERIALS:

The buffer solution consisted of 0.1 mol dm^{-3} THAM and 0.01 mol dm^{-3} acetic acid, adjusted to pH 5.1 with G. F. Smith reagent-grade $HClO_4$. The source of BPh_4^- was a solution of $Ca(BPh_4)_2$ in THAM prepared from Fisher Scientific reagent-grade $NaBPh_4$ by the procedure of Rechnitz et al. (2) and standardized by potentiometric titration with KCl and RbCl. RbCl was from the Fisher Scientific Co.

ESTIMATED ERROR:

Not stated.

Temperature: ±0.3°C

REFERENCES:

1. Howick, L. C.; Pflaum, R. T. *Anal. Chem. Acta* 1958, *19*, 342.
2. Rechnitz, G. A.; Katz, S. A.; Zamochnick, S. B. *Anal. Chem.* 1963, *35*, 1322.

COMPONENTS:

(1) Rubidium tetraphenylborate (1-); $RbC_{24}H_{20}B$; [5971-93-7]

(2) 2-Propanone (acetone); C_3H_6O; [67-64-1]

(3) Water; H_2O; [7732-18-5]

ORIGINAL MEASUREMENTS:

Kirgintsev, A. N.; Kozitskii, V. P. *Izvest. Akad. Nauk SSSR, Khim. Ser.* 1968, 1170-72.

VARIABLES:

Acetone-water composition
One temperature: 25.00°C

PREPARED BY:

Orest Popovych

EXPERIMENTAL VALUES:

The authors reported mass % of $RbBPh_4$ in the saturated solutions, defined as grams of the salt in 100 cm^3 of the solution. The solubilities C have been recalculated to mol dm^{-3} by the compiler.

% Water in acetone* Vol. %	Solubility of $RbBPh_4$ (Wt./vol.)%	10^2C/mol dm^{-3}
0.007	1.56	3.85
2	1.81	4.47
4	1.98	4.89
8	2.10	5.19
12	2.05	5.07
15	1.97	4.87
20	1.73	4.27
25	1.47	3.63
30	1.20	2.97
37	0.75	1.85
45	0.42	1.04
52	0.214	0.529
60	0.113	0.279
70	0.030	0.074
80	0.013	0.032

*Determined by weighing. Solvent volume was taken as the sum of the volumes of acetone and water, neglecting the effect of mixing. The authors provided no density data.

AUXILIARY INFORMATION

METHOD/APPARATUS/PROCEDURE:

Evaporation and weighing.
Saturated solutions were prepared by shaking the suspensions in a constant temperature bath for 6 hours. Aliquots were removed through cotton plugs, weighed and the solvent removed by evaporation first under an IR lamp and then by oven-drying to constant weight at 105°C. The solid phase contained no solvent when recrystallized from acetone or acetone-water mixtures.

SOURCE AND PURITY OF MATERIALS:

$NaBPh_4$ ("analytical grade" from the Apolda Co., GDR) was purified by recrystallization from acetone-toluene, followed by dissolution in water, extraction with ether, and removal of the latter in vacuo. The purity of the final $NaBPh_4$ was no less than 99.6%. $RbBPh_4$ was prepared by metathesis of $NaBPh_4$ with RbCl and purified by double recrystallization from 20% water 80% acetone (probably by vol.). The acetone was slowly evaporated and the precipitate kept for a long time under high vacuum. Acetone was treated with $KMnO_4$ followed by triple fractionation. Final water content was 0.007 vol. % by Karl Fisher titration.

ESTIMATED ERROR:

Precision ±0.5%
Temperature control: ±0.05°C

COMPONENTS:

(1) Rubidium tetraphenylborate (1-); $RbC_{24}H_{20}B$; [5971-93-7]
(2) Acetonitrile; C_2H_3N; [75-05-8]

ORIGINAL MEASUREMENTS:

Popovych, O.; Gibofsky, A.; Berne, D. H. *Anal. Chem.* 1972, *44*, 811-7.

VARIABLES:

One temperature: 25.00°C

PREPARED BY:

Orest Popovych

EXPERIMENTAL VALUES:

The solubility was reported as $C_{BPh_4} = 1.70 \times 10^{-2}$ mol dm^{-3}. The mean molar ionic activity coefficient was calculated using the relationship:

$$-\log y_{\pm} = \frac{1.64C^{\frac{1}{2}}}{1 + 0.485\text{å}C^{\frac{1}{2}}}$$

Adopting å = 0.5 nm for BPh_4^- and å = 0.3 nm for Rb^+, the value of $y_{\pm}^2 = 0.455$ and the pK_{s0}° derived from it was reported as 3.66 (molal scale), i. e., K_{s0}° units are mol^2 kg^{-2}. pK_{s0}° values on the molar scale (K_{s0}° units of mol^2 dm^{-6}) was not reported, but can be calculated from the molal value via the solvent density, which was 0.777 g ml^{-1}. On the molar scale, $K_{s0}^{\circ} = 1.31 \times 10^{-4}$ mol^2 dm^{-6} and the corresponding $pK_{s0}^{\circ} = 3.88$ (compiler). Complete dissociation was assumed, which is generally true for most electrolytes in acetonitrile (1). The molar absorption coefficients for the BPh_4^- ion were determined to be 3203 and 2082 dm^3(cm mol)$^{-1}$ at 266 nm and 274 nm, respectively.

AUXILIARY INFORMATION

METHOD/APPARATUS/PROCEDURE:

UV spectrophotometry using a Cary Model 14 spectrophotometer. Satd solutions were prepd by shaking the suspensions in water-jacketed flasks. After about two weeks of shaking, the suspensions were filtered and the filtrates analyzed spectrophotometrically. The molar absorption coefficients stated above were used to compute the concentration of tetraphenylborate. All work was carried out in deaerated containers and solvents. Differential thermal analysis detected no crystal solvates.

COMMENTS:

The ion-size parameter used for the BPh_4^- ion in this study was probably too small. If the literature value of 1.2 nm (2) is used instead, $y_{\pm}^2$ becomes 0.500, the $K_{s0}^{\circ} = 1.44 \times 10^{-4}$ mol^2 dm^{-6} and $pK_{s0}^{\circ} = 3.84$. Probably a tentative value of $K_{s0}^{\circ} = 1.4 \times 10^{-4}$ mol^2 dm^{-6} can be adopted until the activity coefficients are determined experimentally.

SOURCE AND PURITY OF MATERIALS:

Acetonitrile (Matheson, spectroquality) was refluxed for 24 hrs over CaH_2 and fractionally distilled. $RbBPh_4$ was prepared from $NaBPh_4$ (Fisher, 99.7%) by metathesis with RbCl; it was recrystallized three times from 3:1 acetone-water and dried *in vacuo* at 80°C.

ESTIMATED ERROR:

Precision ±1% (rel.)
Accuracy ±3% (rel.)
Temperature control: ±0.01°C

REFERENCES:

(1) Kay, R. L.; Hales, B. J.; Cunningham, G. P. *J. Phys. Chem.* 1967, *71*, 3925.
(2) Kolthoff, I. M.; Chantooni, M. K. Jr. *Anal. Chem.* 1972, *44*, 194.

COMPONENTS:

(1) Rubidium tetraphenylborate (1-); $RbC_{24}H_{20}B$; [5971-93-7]

(2) 1,2-Dichloroethane; $C_2H_4Cl_2$; [107-06-2]

ORIGINAL MEASUREMENTS:

Abraham, M. H.; Danil de Namor, A. F. *J. Chem. Soc. Faraday Trans. 1* 1976, *72*, 955-62.

VARIABLES:

One temperature: 25°C

PREPARED BY:

Orest Popovych

EXPERIMENTAL VALUES:

The authors reported the solubility of $RbBPh_4$ in 1,2-dichloroethane as:

$$9.90 \times 10^{-6} \text{ mol dm}^{-3}.$$

They used an association constant $K_A = 1.70 \times 10^3$ mol^{-1} dm^3 and the extended Debye-Hückel equation for the mean ionic activity coefficient with an ion-size parameter å = 0.56 nm to calculate the standard Gibbs free energy of solution: $\Delta G_s^\circ = 13.76$ kcal mol^{-1} = 57.60 kJ mol^{-1} (compiler). From the relationship: $\Delta G_s^\circ = -RT \ln K_{s0}^\circ$, the solubility product can be calculated as $pK_{s0}^\circ = 10.088$ where the units of K_{s0}° are mol^2 dm^{-6} (compiler).

AUXILIARY INFORMATION

METHOD/APPARATUS/PROCEDURE:

Saturated solutions prepared by shaking the suspensions at 25°C for several days and analyzing aliquots by evaporation and weighing. Method of temperature control was not specified.

SOURCE AND PURITY OF MATERIALS:

The solvent was shaken with anhydrous K_2CO_3, passed through a column of basic activated alumina into a distillation flask and fractionated under N_2 through a 3-foot column. At least 10% of distillate was rejected, the rest collected over freshly activated molecular sieve. $RbBPh_4$ was recryst. from aqueous acetone and dried in a vacuum oven at 60-80°C for several days.

ESTIMATED ERROR:

Precision of 0.1 kcal mol^{-1} in ΔG_s°.

COMMENTS:

The above solubility product should be regarded as a tentative value.

REFERENCES:

COMPONENTS:

(1) Rubidium tetraphenylborate (1-); $RbC_{24}H_{20}B$; [5971-93-7]
(2) Lithium chloride; LiCl; [7447-41-8]
(3) Ethanol; C_2H_6O; [64-17-5]

ORIGINAL MEASUREMENTS:

1) Popovych, O.; Gibofsky, A.; Berne D. H. *Anal. Chem.* 1972, *44*, 811-7.

2) Berne, D. H. *Ph.D. Thesis*. City University of New York. 1972 (1).

VARIABLES:

LiCl concentration varied from 2 to 200 times that of $RbBPh_4$ in mol dm^{-3}. One Temperature: 25.00°C.

PREPARED BY:

Orest Popovych

EXPERIMENTAL VALUES:

The solubility (ion-activity) product of $RbBPh_4$ in ethanol was reported as:

$$pK^\circ_{s0} = 7.60 \text{ } (K^\circ_{s0} \text{ units are mol}^2 \text{ kg}^{-2}).$$

The value of the ion-activity product, determined from the variation of the solubility with ionic strength, was not reported, but it is listed as $K^\circ_{s0} = 1.56 \times 10^{-8}$ mol^2 dm^{-6} in the Ph.D. thesis by Berne (1).

AUXILIARY INFORMATION

METHOD/APPARATUS/PROCEDURE:

UV spectrophotometry using a Cary Model 14 spectrophotometer. Saturated solutions were prepared by shaking suspensions of $RbBPh_4$ in water-jacketed flasks. After about two weeks of shaking, the suspensions were filtered and the filtrates analyzed spectrophotometrically. The molar absorption coefficients of 2.97×10^3 and 2.10×10^3 dm^3(cm $mol)^{-1}$ at 266 nm and 274 nm, respectively (2), were used to compute the concentration of tetraphenylborate. All work was carried out in deaerated containers and solvents.

SOURCE AND PURITY OF MATERIALS:

$RbBPh_4$ was prepared from $NaBPh_4$ (Fisher, 99.7%) by metathesis with RbCl; it was recrystallized three times from 3:1 acetone-water and dried in vacuo at 80°C. The purification of LiCl and ethanol have been described (3).

ESTIMATED ERROR:

Precision ±1% (rel.) in solubility.
Accuracy ±3% (rel.) in solubility.
Temperature control: ±0.01°C.

REFERENCES:

(1) Berne, D. H. *Ph. D. Thesis.* City University of New York. 1972. (Dissertation index No. 73-02829).
(2) Dill, A. J.;Popovych, O. *J. Chem. Eng. Data* 1969, *14*, 240.
(3) Dill, A. J.; Popovych, O. *J. Chem. Eng. Data* 1969, *14*, 240.

COMPONENTS:

(1) Cesium tetraphenylborate (1-); $CsC_{24}H_{20}B$; [3087-82-9]

(2) Water; H_2O; [7732-18-5]

EVALUATOR:

Orest Popovych, Department of Chemistry, City University of New York, Brooklyn College, Brooklyn, N. Y. 11210, U. S. A.
October 1979

CRITICAL EVALUATION:

There are seven publications dealing with original data pertaining to the solubility of cesium tetraphenylborate ($CsBPh_4$) in aqueous solutions (1-7). Two of them, however, report only the solubility product (1,2). Only one of the studies deals with the solubility as a function of the temperature, ionic strength and pH (3). In one publication (7) the solubility was reported in a buffer solution,but not in pure water. What seems to be historically the first datum, the solubility product of $CsBPh_4$ reported as 5 x 10^{-10} (presumably mol^2 dm^{-6}) at 290 K by Rüdorff and Zannier (1), must be rejected because of lack of any experimental details provided, except for the statement that the determination was conductometric. Since the result was reported to one significant figure, no compilation sheet was provided for that original source.

Solubility at 298 K.

At 298 K, we have the solubility data of Pflaum and Howick (4), Alexander and Parker (2), expressed in the form of a solubility product, and of Popovych, Gibofsky and Berne (5). All three determinations were by UV-spectrophotometry for the BPh_4^- ion, but they differ in their reported degree of temperature control and saturation control. Alexander and Parker in reporting a pK_{s0} of 8.7 (all solubility products in this evaluation have units of mol^2 dm^{-6}) failed to specify how the temperature was controlled and how the existence of saturation was ascertained. However, if we make the usual assumption that the error is 0.1 pK units, the solubility derived from the above K_{s0} by simply taking the square root is (4.5 ± 0.5) x 10^{-5} mol dm^{-3}. This calculation is justified since the above solubility product was reported to be a product of concentrations, not activities. As we can see, the solubility value derived from the datum of Alexander and Parker does not agree with the value of 2.79 x 10^{-5} mol dm^{-3} reported by Pflaum and Howick. In the latter study, there is also no indication of how the temperature and the saturation were controlled. Furthermore, the authors used the molar absorption coefficients determined in acetonitrile to analyze the spectra obtained on aqueous solutions. While this procedure may have resulted in a compensation of errors (see evaluation for $KBPh_4$ in water), a combination of all those shortcomings cannot inspire confidence in the reported solubility value. The work of Popovych et al. (5), on the other hand, was characterized by temperature control to ±0.01°C and repeated analyses days apart until a constant solubility value was attained. The relative precision of determining the solubility of $CsBPh_4$ in water turned out to be a little lower than the normal ±1% associated with the UV-analysis for tetraphenylborates -- it was ±4%. The solubility was reported as 4.0 x 10^{-5} mol dm^{-3} (the last, uncertain, digit dropped by the evaluator). If we restirct the recommendation to two significant digits in the solubility value, it may be just barely justified to average the value of Alexander and Parker (2) and that of Popovych et al. (5) to arrive at the recommended solubility value at 298 K as:

$$\text{Solubility} = (4.2 \pm 0.2) \times 10^{-5} \text{ mol dm}^{-3}.$$

The solubility product calculated as $K^\circ_{s0} = C^2 y_\pm^2$, where C is the solubility and $y_\pm^2$ is the mean ionic activity coefficient calculated from the Debye-Hückel limiting law, was reported by Popovych et al. (5) as pK°_{s0} = 8.80. It is within 0.1 log units of the value reported by Alexander and Parker (2), but the latter contains no activity correction.

Solubility at Other Temperatures

There are two reported solubility values at 293 K. From the work of Geilman and Gebauhr (6), the solubility calculated by the compiler is 3.2 x 10^{-5} mol dm^{-3}. At zero ionic strength (presumably meaning in the absence of added inert eletrolyte) the solubility value reported by Siska (3) is 3.50 x 10^{-5} mol dm^{-3}. Although one might be tempted to average

COMPONENTS:

(1) Cesium tetraphenylborate (1-);

$CsC_{24}H_{20}B$; [3087-82-9]

(2) Water; H_2O; [7732-18-5]

EVALUATOR:

Orest Popovych, Department of Chemistry, City University of New York, Brooklyn College, Brooklyn, N. Y. 11210, U. S. A.
October 1979

CRITICAL EVALUATION:

these two results at 293 K, the problem with Siska's data is the very short time of equilibration (3 hours). This length of equilibration proved to be sufficient for the saturation of $CsBPh_4$ in the rate-of-dissolution study by Geilman and Gebauhr (6), but it is not so as a general rule, in the experience of this evaluator. We have noted in the evaluation for $KBPh_4$ in water that Siska's (3) solubility values seemed to be too low. If undersaturation did not occur for $CsBPh_4$, much valuable information might be salvaged, because Siska's is the only study to date that reported the solubility of $CsBPh_4$ in water as a function of ionic strength and pH at 293 K as well as the solubility at other temperatures in solutions with ionic strength maintained at 0.1 mol dm^{-3}. The latter solubility determinations were reported at 283 K, 303 K, 313 K and 318 K (3). Unfortunately, the only other literature datum on the solubility of $CsBPh_4$ in a 0.1 mol dm^{-3} buffer solution, which is 5.4×10^{-5} mol dm^{-3} at 298 K (7), tends to confirm the undersaturation of Siska's solution.

REFERENCES:

1. Rüdorff, W.; Zannier, H. *Angew. Chem.* 1952, *64*, 613.
2. Alexander, R.; Parker, A. J. *J. Am. Chem. Soc.* 1967, *89*, 5549.
3. Siska, E. *Magy. Kem. Foly.* 1976, *82*, 275.
4. Pflaum, R. T.; Howick, L. C. *Anal. Chem.* 1956, *28*, 1542.
5. Popovych, O; Gibofsky, A.; Berne, D. H. *Anal. Chem.* 1972, *44*, 811.
6. Geilman, W.; Gebauhr, W. *Z. anal. Chem.* 1953, *139*, 161.
7. McClure, J. E.; Rechnitz, G. A. *Anal. Chem.* 1966, *38*, 136.

COMPONENTS:

(1) Cesium tetraphenylborate (1-); $CsC_{24}H_{20}B$; [3087-82-9]

(2) Water; H_2O; [7732-18-5]

ORIGINAL MEASUREMENTS:

Alexander, R.; Parker, A. J.
J. Am. Chem. Soc. 1967, *89*, 5549-51.

VARIABLES:

One temperature: 25°C

PREPARED BY:

Orest Popovych

EXPERIMENTAL VALUES:

The formal (concentration) solubility product of $CsBPh_4$ in water was reported as:

$$pK_{s0} = 8.7 \ (K_{s0} \text{ units are mol}^2 \text{ dm}^{-6}).$$

The solubility can be calculated as $(K_{s0})^{1/2} = (4.5 \pm 0.5) \times 10^{-5}$ mol dm^{-3} (compiler).

AUXILIARY INFORMATION

METHOD/APPARATUS/PROCEDURE:

UV spectrophotometry on solutions saturated under nitrogen. No other details.

SOURCE AND PURITY OF MATERIALS:

Not stated.

ESTIMATED ERROR:

Nothing specified.
Precision of ±0.1 pK is assumed by the compiler.

REFERENCES:

COMPONENTS:

(1) Cesium tetraphenylborate (1-); $CsC_{24}H_{20}B$; [3087-82-9]

(2) Sodium sulfate; Na_2SO_4; [7757-82-6]

(3) Water; H_2O; [7732-18-5]

ORIGINAL MEASUREMENTS:

Siska, E. *Magy. Kem. Foly.* 1976, *82*, 275-8.

VARIABLES:

Temperature range 10-45°C
Concentration of Na_2SO_4
pH

PREPARED BY:

Orest Popovych

EXPERIMENTAL VALUES:

In distilled water at 20°C, the solubility of $CsBPh_4$ was reported to be: 3.49×10^{-5} mol dm^{-3} and the corresponding solubility product, K_{s0}, as 1.22×10^{-9} mol^2 dm^{-6}. The K_{s0} is simply the square of the solubility, without activity corrections.

With ionic strength varied by means of Na_2SO_4, the following solubilities, C, were reported for $CsBPh_4$ in aqueous solution at 20°C:

Ionic strength/mol dm^{-3}	10^5C/mol dm^{-3}
0	3.50
0.05	3.00
0.1	2.98
0.3	2.55
0.5	2.35
0.7	2.03
1.0	1.58
2.0	0.86

AUXILIARY INFORMATION

METHOD/APPARATUS/PROCEDURE:

Amperometric titration. For details see compilation for $KBPh_4$ in water based on the same reference. $CsBPh_4$ was prepared from the chloride by metathesis with $NaBPh_4$.

SOURCE AND PURITY OF MATERIALS:

Not specified.

ESTIMATED ERROR:

Precision in solubility determination is ±2%.
Temperature control: ±1°C.

REFERENCES:

COMPONENTS:

(1) Cesium tetraphenylborate (1-); $CsC_{24}H_{20}B$; [3087-82-9]

(2) Sodium sulfate; Na_2SO_4; [7757-82-6]

(3) Water; H_2O; [7732-18-5]

ORIGINAL MEASUREMENTS:

Siska, E. *Magy. Kem. Foly.* 1976, *82*, 275-8.

VARIABLES:

PREPARED BY:

COMMENTS AND/OR ADDITIONAL DATA

EXPERIMENTAL VALUES:

Keeping the ionic strength constant at 0.1 mol dm^{-3} with Na_2SO_4, the following solubilities of $CsBPh_4$, C, were obtained as a function of the temperature:

t/°C	10^5C/mol dm^{-3}
10	2.02
20	2.80
30	3.90
40	5.72
45	6..80

Keeping the ionic strength constant at 0.1 mol dm^{-3} with Na_2SO_4, the following solubilities of $CsBPh_4$, C, were obtained at 20°C as a function of pH varied by means of acetic acid and sodium hydroxide:

pH	10^5C/mol dm^{-3}
1.7	1.48
2.3	2.52
2.7	2.74
2.9	2.72
4.4	2.78
5.4	2.85
6.5	2.79
7.3	2.87
7.9	2.86
8.5	2.84
9.3	2.82
10.1	2.82
10.9	2.87
11.3	2.82

At 20°C, ionic strength of 0.1 mol dm^{-3} and in the pH range of 2.7-11.3, the authors report the solubility of $CsBPh_4$ in water as:

$C = (2.82 \pm 0.047) \times 10^{-5}$ mol dm^{-3}.

AUXILIARY INFORMATION

METHOD/APPARATUS/PROCEDURE:

SOURCE AND PURITY OF MATERIALS:

ESTIMATED ERROR:

REFERENCES:

COMPONENTS:

(1) Cesium tetraphenylborate (1-); $CsC_{24}H_{20}B$; [3087-82-9]

(2) Water: H_2O; [7732-18-5]

ORIGINAL MEASUREMENTS:

Pflaum, R. T.; Howick, L. C.
Anal. Chem. 1956, *28*, 1542-4.

VARIABLES:

One temperature: 25°C

PREPARED BY:

Orest Popovych

EXPERIMENTAL VALUES:

The solubility of $CsBPh_4$ in water was reported as 2.79×10^{-5} mol dm^{-3}.

AUXILIARY INFORMATION

METHOD/APPARATUS/PROCEDURE:

Ultraviolet spectrophotometry. For details see the compilation for $KBPh_4$ in water based on the same reference.

SOURCE AND PURITY OF MATERIALS:

See the compilation for $KBPh_4$ in water based on the same reference. $CsBPh_4$ was prepared by metathesis of CsCl and $NaBPh_4$ and recrystallized from an acetonitrile-water mixture.

ESTIMATED ERROR:

Nothing is specified, but the precision is likely to be ±1% (compiler).

REFERENCES:

COMPONENTS:

(1) Cesium tetraphenylborate (1-); $CsC_{24}H_{20}B$; [3087-82-9]

(2) Water; H_2O; [7732-18-5]

ORIGINAL MEASUREMENTS:

Popovych, O.; Gibofsky, A.; Berne, D. H. *Anal. Chem.* 1972, *44*, 811-7.

VARIABLES:

One temperature: 25.00°C

PREPARED BY:

Orest Popovych

EXPERIMENTAL VALUES:

The solubility of $CsBPh_4$ was reported as: $C_{BPh_4} = 4.0_1 \times 10^{-5}$ mol dm^{-3}.

Combining the above value with the mean molar activity coefficient calculated from the Debye-Hückel limiting law $-\log y_{\pm} = 0.509C^{\frac{1}{2}}$, the authors reported as the solubility product: $pK^{\circ}_{s0} = 8.80$ (K°_{s0} units are mol^2 kg^{-2}).

AUXILIARY INFORMATION

METHOD/APPARATUS/PROCEDURE:

Ultraviolet spectrophotometry. Procedure identical with that described for $KBPh_4$ in methanol. Differential thermal analysis showed absence of crystal solvates.

SOURCE AND PURITY OF MATERIALS:

$CsBPh_4$ was prepared and purified by a method analogous to that employed for $KBPh_4$ and described in the compilation for $KBPh_4$ in methanol.

ESTIMATED ERROR:

Precision ±4% (rel.)
Temperature control: ±0.01°C

REFERENCES:

COMPONENTS:

(1) Cesium tetraphenylborate (1-); $CsC_{24}H_{20}B$; [3087-82-9]

(2) Water; H_2O; [7732-18-5]

ORIGINAL MEASUREMENTS:

Geilman, W.; Gebauhr, W.
Z. anal. Chem. 1953, *139*, 161-81.

VARIABLES:

One temperature: 20°C

PREPARED BY:

Orest Popovych

EXPERIMENTAL VALUES:

The solubility is reported both as c_{Cs} = 0.430 mg cm^{-3} and as 2.9 x 10^{-5} mol dm^{-3}. The corresponding solubility product, calculated simply as c_{Cs}^2, is reported as 8.4 x 10^{-10}.

However, inspection of the raw data (below) suggests that the authors consider only two figures to be significant, i. e., the solubility should have been reported as 0.43 mg cm^{-3}. Using the value of 132.91 for the relative atomic mass of Cs, this compiler obtains somewhat different results from the authors:

c_{Cs} = 3.2 x 10^{-5} mol dm^{-3} and $K_{s0} \equiv c_{Cs}^2$ = 1.0 x 10^{-9} mol^2 dm^{-6}.

Also reported is the rate of dissolution of $CsBPh_4$ in water:

Time, hours	μg Cs/10 cm^3 of water
0.5	28.0
1.0	34.0
3.0	44.0
6.0	43.5
12.0	43.5
20.0	43.2

AUXILIARY INFORMATION

METHOD/APPARATUS/PROCEDURE:

Radiometric, using liquid-scintillation counting of ^{134}Cs. The radioactive cesium obtained as the carbonate from the Harwell nuclear reactor was purified by precipitation with $HClO_4$ in the presence of 2 mg of $NaHPO_4$ followed by recrystallization. The $CsClO_4$ solution was reacted with $NaBPh_4$, the resulting $CsBPh_4$ precipitate washed with water, mechanically shaken in water at 20°C and the filtrate analyzed radiometrically to constant activity. Apparatus was not specified.

SOURCE AND PURITY OF MATERIALS:

Not stated.

ESTIMATED ERROR:

Not specified. However, given the temperature control to ±0.5°C, the relative precision cannot be better than ±1-2%.

REFERENCES:

COMPONENTS:

(1) Cesium tetraphenylborate (1-); $CsC_{24}H_{20}B$ [3087-82-9]
(2) Tris(hydroxymethyl)aminomethane $C_4H_{11}NO_3$; [77-86-1]
(3) Acetic acid; $C_2H_4O_2$; [64-19-7]
(4) Water; H_2O; [7732-18-5]

ORIGINAL MEASUREMENTS:

McClure, J. E.; Rechnitz, G. A. *Anal. Chem.* 1966, *38*, 136-9.

VARIABLES:

One temperature: 24.8°C

PREPARED BY:

Orest Popovych

EXPERIMENTAL VALUES:

The solubility of cesium tetraphenylborate ($CsBPh_4$) in aqueous tris(hydroxymethyl)aminomethane (THAM) buffer at pH 5.1 was reported as:

$$5.4 \times 10^{-5} \text{ mol dm}^{-3}.$$

AUXILIARY INFORMATION

METHOD/APPARATUS/PROCEDURE:

UV-spectrophotometry according to the procedure of Howick and Pflaum (1).
No other details.

SOURCE AND PURITY OF MATERIALS:

The buffer solution consisted of 0.1 mol dm^{-3} THAM and 0.01 mol dm^{-3} acetic acid, adjusted to pH 5.1 with G. F. Smith reagent-grade $HClO_4$. The source of BPh_4^- was a solution of $Ca(BPh_4)_2$ in THAM prepared from Fisher Scientific reagent-grade $NaBPh_4$ by the procedure of Rechnitz et al. (2) and standardized by potentiometric titration with KCl and RbCl. CsCl was from the Fisher Scientific Company.

ESTIMATED ERROR:

Not stated.
Temperature: ±0.3°C

REFERENCES:

1) Howick, L. C.; Pflaum, R. T. *Anal. Chem. Acta* 1958, *19*, 342.
2) Rechnitz, G. A.; Katz, S. A.; Zamochnick, S. B. *Anal. Chem.* 1963, *35*, 1322.

COMPONENTS:
(1) Cesium tetraphenylborate (1-); $CsC_{24}H_{20}B$; [3087-82-9]
(2) Lithium chloride; LiCl; [7447-41-8]
(3) Sodium hydroxide; NaOH; [1310-73-2]
(4) Methanol; CH_4O; [67-56-1]
(5) Water; H_2O; [7732-18-5]

ORIGINAL MEASUREMENTS:
Berne, A; *M. A. Thesis*. Brooklyn College. 1976.

VARIABLES:
Methanol-water composition. LiCl concentration varied from 0 to 2,000 times the solubility of $CsBPh_4$.
One temperature: 25.00°C

PREPARED BY:
Orest Popovych

EXPERIMENTAL VALUES:

The author reported the mean ionic activities of $CsBPh_4$ in saturated solutions, which were determined from the variation of the solubility of $CsBPh_4$ as a function of ionic strength varied by means of LiCl in the presence of 5 x 10^{-5} mol dm^{-3} NaOH. For each solvent composition, the activity in saturated solution without added inert electrolyte, $a_{\pm,o}$, was computed from the functions:

$$\log C_I = \log a_{\pm,o} - \log y_{\pm,I} \quad \text{and} \quad -\log y_{\pm,I} \doteq A_1 I^{1/2} + A_2 I + A_3 I^{3/2} + \ldots$$

where $y_{\pm}$ is the mean molar activity coefficient ($f_{\pm}$ in the original) and the subscripts o and I denote solutions without and with added inert electrolyte. Complete dissociation was assumed. The solubilities at different ionic strength were not reported, but their values, as well as those of the activity coefficients at any ionic strength I (in mol dm^{-3}) can be calculated from the A-coefficients characteristic of the methanol-water mixture tabulated below.

The solubility of $CsBPh_4$ in the pure solvents was not measured due to the extensive decomposition in the absence of added LiCl, but the corresponding ionic activity, $a_{\pm,o}$, should be a very good approximation of the solubility in the concentration range involved.

continued ....

AUXILIARY INFORMATION

METHOD/APPARATUS/PROCEDURE:
Ultraviolet spectrophotometry using a Cary Model 17 spectrophotometer. Saturation was achieved by subjecting the suspensions for 20 minutes to ultrasonic waves (E/MC Corp. Model 450 Ultrasonic Generator) followed by agitation in a thermostatted bath for at least 3 days until constant spectral absorption. Spectra were monitored carefully for indications of decomposition, to which $CsBPh_4$ solutions are highly susceptible. Sodium hydroxide was added to retard the decomposition.

SOURCE AND PURITY OF MATERIALS:
$CsBPh_4$ was prepared by metathesis of $NaBPh_4$ and CsCl (both from Alfa Inorganic) and purified by triple recrystallization from a 3:1 acetone-water mixture. It was dried in vacuo at 80°C for 24 hours. LiCl (Baker Analyzed Reagent) was used without purification after drying in vacuo at 110°C for 24 hours. Certified ACS spectroanalyzed methanol (Fisher Scientific Co.) was used as received. The mass% of methanol-water mixtures was determined from the measured densities and literature data (1).

ESTIMATED ERROR:
The relative precision of $a_{\pm,o}$ is tabulated.
Temperature control: ±0.01°C.

REFERENCES:

(1) Bates, R. G.; Robinson, R. A. in *Chemical Physics of Ionic Solutions*, Conway, B. E.; Barradas, R. G., Eds. Wiley. New York. 1966. Chapter 12.

COMPONENTS:	EVALUATOR:
(1) Cesium tetraphenylborate (1-); $CsC_{24}H_{20}B$; [3087-82-9] (2) Lithium chloride; LiCl; [7447-41-8] (3) Sodium hydroxide; NaOH; [1310-73-2] (4) Methanol; CH_4O; [67-56-1] (5) Water; H_2O; [7732-18-5]	Orest Popovych, Department of Chemistry, The City University of New York, Brooklyn College, Brooklyn, N. Y. 11210, U. S. A. September]979

CRITICAL EVALUATION:

COMMENTS AND/OR ADDITIONAL DATA

Mass% methanol in water	$10^4 a_{\pm,o}$/mol dm^{-3}	Relative error, % $da_{\pm,o}/a_{\pm,o}$	A_1	A_2	A_3	A_4
100.0 a	5.04	3.6	1.81	-3.59	4.22	-2.17
89.1 b	4.43	2.0	1.22	-1.81	1.49	-0.564
79.8	4.07	2.4	1.03	-1.18	0.487	----
70.7	3.34	5.4	1.16	-1.89	1.34	----
61.1	3.09	7.0	0.567	-0.440	0.0864	----
49.9	1.83	21	0.747	-0.177	-3.91	4.45
39.9	1.22	6.2	0.661	-0.878	0.432	----
29.5	0.749	5.8	0.630	-0.698	0.167	----
20.1	0.474	4.0	0.559	-0.934	0.660	----

a No NaOH was added to solutions in 100.0% methanol.

b The data in this row were recalculated by the compiler using a more reliable value for the absorption coefficient for the tetraphenylborate ion. In the original,
$a_{\pm,o} = 4.49 \times 10^{-4}$ mol dm^{-3}; $A_1 = 1.25$; $A_2 = -1.97$; $A_3 = 1.77$; and $A_4 = -0.724$.

COMPONENTS:

(1) Cesium tetraphenylborate (1-); $CsC_{24}H_{20}B$; [3087-82-9]

(2) 2-Propanone (acetone); C_3H_6O; [67-64-1]

(3) Water; H_2O; [7732-18-5]

ORIGINAL MEASUREMENTS:

Kirgintsev, A. N.; Kozitskii, V. P. *Izvest. Akad. Nauk SSSR, Khim. Ser.* 1968, 1170-2.

VARIABLES:

Acetone-water composition
One temperature: 25.00°C

PREPARED BY:

Orest Popovych

EXPERIMENTAL VALUES:

The authors reported mass % of $CsBPh_4$ in the saturated solutions, defined as grams of salt in 100 cm^3 of the solution. The solubilities have been recalculated to mol dm^{-3} by the compiler.

Vol. % water in acetone*	Solubility of $CsBPh_4$ (Wt./vol.)%	10^2C/mol dm^{-3}
0.007	1.50	3.32
2	1.69	3.74
4	1.82	4.03
8	1.92	4.25
12	1.88	4.16
15	1.80	3.98
20	1.56	3.45
25	1.32	2.92
30	1.05	2.32
37	0.71	1.57
45	0.38	0.84
52	0.20	0.44
60	0.077	0.170
70	0.034	0.075
80	0.011	0.024

*Determined by weighing. Solvent volume was taken as the sum of the volumes of acetone and water, neglecting the effect of mixing.

AUXILIARY INFORMATION

METHOD/APPARATUS/PROCEDURE:

Evaporation and weighing. For details see compilation for $KBPh_4$ in acetone-water, based on the above reference.

SOURCE AND PURITY OF MATERIALS:

See compilation for $KBPh_4$ in acetone-water based on the above reference. $CsBPh_4$ was prepared and purified as $KBPh_4$, starting with CsCl.

ESTIMATED ERROR:

Precision ±0.5%
Temperature control: ±0.05°C

REFERENCES:

COMPONENTS:

(1) Cesium tetraphenylborate (1-); $CsC_{24}H_{20}B$; [3087-82-9]

(2) Acetonitrile; C_2H_3N; [75-05-8]

EVALUATOR:

Orest Popovych, Department of Chemistry, City University of New York, Brooklyn College, Brooklyn, N. Y. 11210, U. S. A.
October 1979

CRITICAL EVALUATION:

There are only two original data on the solubility of cesium tetraphenylborate ($CsBPh_4$) in acetonitrile, both at 298 K. Alexander and Parker (1) reported the solubility in the form pK_{s0} = 3.1, where the K_{s0} was a product of ionic concentrations, not activities (all K_{s0} values in this evaluation have units of $mol^2\ dm^{-6}$). On the assumption that the above pK_{s0} has a precision of ±0.1 units, the compiler estimated the solubility to be $(K_{s0})^{1/2}$ = (2.8 ± 0.3) x 10^{-2} mol dm^{-3}. The analytical method was either UV-spectrophotometry, or titration of BPh_4^- with Ag^+. Unfortunately, no experimental details are provided by the authors as to the extent of temperature control and the manner in which the attainment of saturation was ascertained. Popovych, Gibofsky and Berne (2) reported a solubility value of 1.68 x 10^{-2} mol dm^{-3}, which differs greatly from that of Alexander and Parker (1). Considering that Popovych et al. (2) controlled the temperature of the bath to 0.01°C and that saturation (as well as possible decomposition) were monitored by successive analyses days apart until the results checked to 1% or better, their solubility value would seem to be the preferred one of the two. However, the <u>solubility of 1.68 x 10^{-2} mol dm^{-3}</u> should be considered no better than <u>tentative</u> at this time.

The thermodynamic solubility product estimated by Popovych et al. (2) was reported in the form pK_{s0}° = 3.67 (weight basis), i. e., K_{s0}° units of $mol^2\ kg^{-2}$. Based on volume units, i. e., mol dm^{-3}, the pK_{s0}° would be 3.89 (evaluator). The activity correction was made via the calculated activity coefficient of $y_{\pm}^2$ = 0.456 derived from the Debye-Hückel equation with ion-size parameter shown on the compilation sheet. One could argue, however, that using $\mathring{a}$ = 0.5nm for the BPh_4^- ion is not realistic; for example, Kolthoff and Chantooni (3) preferred a value of $\mathring{a}$ = 1.2nm. Using the latter value in our calculation leads to a $y_{\pm}^2$ = 0.501 and a pK_{s0}° = 3.85 (volume basis).

REFERENCES:

1. Alexander, R.; Parker, A. J. *J. Am. Chem. Soc.* 1967, *89*, 5549.
2. Popovych, O.; Gibofsky, A.; Berne, D. H. *Anal. Chem.* 1972, *44*, 811.
3. Kolthoff, I. M.; Chantooni, M. K., Jr. *Anal. Chem.* 1972, *44*, 194.

COMPONENTS:

(1) Cesium tetraphenylborate (1-); $CsC_{24}H_{20}B$; [3087-82-9]

(2) Acetonitrile; C_2H_3N; [75-05-8]

ORIGINAL MEASUREMENTS:

Alexander, R.; Parker, A. J.
J. Am. Chem. Soc. 1967, *89*, 5549-51.

VARIABLES:

One Temperature: 25°C

PREPARED BY:

Orest Popovych

EXPERIMENTAL VALUES:

The formal (concentration) solubility product of $CsBPh_4$ in acetonitrile was reported as :

$$pK_{s0} = 3.1 \ (K_{s0} \text{ units are } mol^2 \ dm^{-6}).$$

The solubility is therefore $(K_{s0})^{1/2} = (2.8 \pm 0.3) \times 10^{-2}$ mol dm^{-3} (compiler).

AUXILIARY INFORMATION

METHOD/APPARATUS/PROCEDURE:

UV spectrophotometry on solutions saturated under nitrogen, or titration for the BPh_4^- anion with silver nitrate. No other details.

SOURCE AND PURITY OF MATERIALS:

Not stated.

ESTIMATED ERROR:

Nothing is specified.
A precision of ±0.1 pK units can be assumed (compiler).

REFERENCES:

COMPONENTS:

(1) Cesium tetraphenylborate (1-); $CsC_{24}H_{20}B$; [3087-82-9]

(2) Acetonitrile; C_2H_3N; [75-05-8]

ORIGINAL MEASUREMENTS:

Popovych, O.; Gibofsky, A.; Berne, D. H. *Anal. Chem.* 1972, *44*, 811-7.

VARIABLES:

One temperature: 25.00°C

PREPARED BY:

Orest Popovych

EXPERIMENTAL VALUES:

The solubility was reported as $C_{BPh_4} = 1.68 \times 10^{-2}$ mol dm^{-3}.

The mean molar ionic activity coefficient was calculated using the relationship:

$$-\log y_{\pm} = \frac{1.64\ C^{1/2}}{1 + 0.485\mathring{a}C^{1/2}}$$

Adopting $\mathring{a}$ = 0.5 nm for BPh_4^- and $\mathring{a}$ = 0.3 nm for Cs^+, the value of $y_{\pm}^2$ = 0.456, and the pK°_{s0} = 3.67 (weight basis), i. e., K°_{s0} units are mol^2 kg^{-2}. pK°_{s0} value on the volume basis (K°_{s0} units of mol^2 dm^{-6}) was not reported, but can be calculated from the above value via the solvent density, which was 0.777 g ml^{-1}. Complete dissociation was assumed, which is generally true for most electrolytes in acetonitrile.

AUXILIARY INFORMATION

METHOD/APPARATUS/PROCEDURE:

Ultraviolet spectrophotometry. Procedure identical with that described for $KBPh_4$ in methanol. Differential thermal analysis showed absence of crystal solvates.

SOURCE AND PURITY OF MATERIALS:

Acetonitrile (Matheson, spectroquality) was refluxed for 24 hours over CaH_2 and fractionally distilled. $CsBPh_4$ was prepared and purified in a manner analogous to that described in the compilation for $KBPh_4$ in methanol.

ESTIMATED ERROR:

Precision ±1% (rel.)
Accuracy ±3% (rel.)
Temperature control: 0.01°C

REFERENCES:

COMPONENTS:

(1) Cesium tetraphenylborate (1-); $CsC_{24}H_{20}B$; [3087-82-9]

(2) 1,1-Dichloroethane; $C_2H_4Cl_2$; [75-34-3]

ORIGINAL MEASUREMENTS:

Abraham, M. H.; Danil de Namor, A. F. *J. Chem. Soc. Faraday Trans. 1* 1976, *72*, 955-62.

VARIABLES:

One temperature: 25°C

PREPARED BY:

Orest Popovych

EXPERIMENTAL VALUES:

The authors reported the solubility of $CsBPh_4$ in 1,1-dichloroethane as:

$$5.30 \times 10^{-5} \text{ mol dm}^{-3}.$$

Using an estimated association constant of 1.20×10^4 mol^{-1} dm^3 and an ion-size parameter of å = 0.57 nm with which to calculate the mean ionic activity coefficient from the extended Debye-Hückel equation, they obtained for the standard Gibbs free energy of solution:

$$\Delta G_s^\circ = 12.21 \text{ kcal mol}^{-1} = 51.11 \text{ kJ mol}^{-1} \text{ (compiler).}$$

The solubility (ion-activity) product of $CsBPh_4$ can be calculated from the relationship:

$\Delta G_s^\circ = -RT \ln K_{s0}^\circ$, yielding $pK_{s0}^\circ = 8.952$, where K_{s0}° units are mol^2 dm^{-6} (compiler).

AUXILIARY INFORMATION

METHOD/APPARATUS/PROCEDURE:

Evaporation and weighing. Saturated solutions prepared by shaking the suspensions for several days at 25°C. No solvate was detected. Method of temperature control was not specified.

SOURCE AND PURITY OF MATERIALS:

The solvent was shaken with anhydrous K_2CO_3, passed through a column of basic activated alumina into a distillation flask and fractionated under N_2 through a three-foot column. At least 10% of distillate was rejected, the rest collected over freshly activated molecular sieve. $CsBPh_4$ was recrystallized from aqueous acetone and vacuum dried at 60-80°C for several days.

ESTIMATED ERROR:

Precision of 0.1 kcal mol^{-1} in ΔG_s°.

REFERENCES:

COMPONENTS:

(1) Cesium tetraphenylborate (1-); $CsC_{24}H_{20}B$; [3087-82-9]

(2) 1,2-Dichloroethane; $C_2H_4Cl_2$; [107-06-2]

ORIGINAL MEASUREMENTS:

Abraham, M. H.; Danil de Namor, A. F. *J. Chem. Soc. Faraday Trans. 1* 1976, *72*, 955-62.

VARIABLES:

One temperature: 25°C

PREPARED BY:

Orest Popovych

EXPERIMENTAL VALUES:

The authors reported the solubility of $CsBPh_4$ in 1,2-dichloroethane as:

$$3.09 \times 10^{-5} \text{ mol dm}^{-3}.$$

Using an estimated association constant of 2.00×10^3 mol^{-1} dm^3 and an ion-size parameter of å = 0.57 nm with which to calculate the mean ionic activity coefficient from the extended Debye-Hückel equation, they arrived at the value for the standard Gibbs free energy of solution:

$$\Delta G^\circ_s = 12.51 \text{ kcal mol}^{-1} = 52.37 \text{ kJ mol}^{-1} \text{ (compiler)}.$$

The solubility (ion-activity) product of $CsBPh_4$ can be calculated from the relationship:

$\Delta G^\circ_s = -RT \ln K^\circ_{s0}$, yielding $pK^\circ_{s0} = 9.172$, where K°_{s0} units are mol^2 dm^{-6} (compiler).

AUXILIARY INFORMATION

METHOD/APPARATUS/PROCEDURE:

Evaporation and weighing. Saturated solutions prepared by shaking the suspensions for several days at 25°C. No solvate was detected. Method of temperature control was not specified.

SOURCE AND PURITY OF MATERIALS:

The solvent was shaken with anhydrous K_2CO_3, passed through a column of basic activated alumina into a distillation flask and fractionated under N_2 through a three foot column. At least 10% of the distillate was rejected, the rest collected over freshly activated molecular sieve. $CsBPh_4$ was recrystallized from aqueous acetone and vacuum dried at 60-80°C for several days.

ESTIMATED ERROR:

Precision of 0.1 kcal mol^{-1} in ΔG°_s.

REFERENCES:

COMPONENTS:

(1) Cesium tetraphenylborate (1-); $CsC_{24}H_{20}B$; [3087-82-9]

(2) Lithium chloride; LiCl; [7447-41-8]

(3) Ethanol; C_2H_6O; [64-17-5]

ORIGINAL MEASUREMENTS:

Popovych, O.; Gibofsky, A.; Berne, D. H. *Anal. Chem.* 1972, *44*, 811-7.

VARIABLES:

LiCl concentration varied from 2 to 200 times that of $CsBPh_4$ in mol dm^{-3}. One temperature: 25.00°C.

PREPARED BY:

Orest Popovych

EXPERIMENTAL VALUES:

The solubility (ion-activity) product of $CsBPh_4$ in ethanol was reported as:

$$pK^{\circ}_{s0} = 7.65 \ (K^{\circ}_{s0} \text{ units are mol}^2 \text{ kg}^{-2}).$$

The value of the ionic activity, determined from the variation of the solubility with ionic strength, was not reported.

AUXILIARY INFORMATION

METHOD/APPARATUS/PROCEDURE:

UV spectrophotometry. Procedure identical with that described for $KBPh_4$ in methanol. Differential thermal analysis showed absence of crystal solvates.

COMMENTS:

$CsBPh_4$ is susceptible to decomposition in solution.

SOURCE AND PURITY OF MATERIALS:

$CsBPh_4$ was prepared and purified by a method analogous to that employed for $KBPh_4$ and described in the compilation for $KBPh_4$ in methanol. The purification of LiCl and ethanol have been described (1).

ESTIMATED ERROR:

Precision ±1% (rel.) in solubility.
Accuracy ±3% (rel.) in solubility.
Temperature control: ±0.01°C

REFERENCES:

(1) Dill, A. J.; Popovych, O. *J. Chem. Eng. Data* 1969, *14*, 240.

COMPONENTS:

(1) Cesium tetraphenylborate (1-); $CsC_{24}H_{20}B$; [3087-82-9]

(2) Formamide; CH_3NO; [75-12-7]

ORIGINAL MEASUREMENTS:

Alexander, R.; Parker, A. J.
J. Am. Chem. Soc. 1967, *89*, 5549-51.

VARIABLES:

One temperature: 25°C

PREPARED BY:

Orest Popovych

EXPERIMENTAL VALUES:

The formal (concentration) solubility product of $CsBPh_4$ in formamide was reported as:

$$pK_{s0} = 3.6 \ (K_{s0} \text{ units are mol}^2 \text{ dm}^{-6}).$$

The solubility can be estimated as $(K_{s0})^{1/2} = (1.6 \pm 0.2) \times 10^{-2}$ mol dm^{-3} (compiler).

AUXILIARY INFORMATION

METHOD/APPARATUS/PROCEDURE:

UV spectrophotometry on solutions saturated under nitrogen, or titration for the BPh_4^- anion with silver nitrate. No other details.

SOURCE AND PURITY OF MATERIALS:

Not stated.

ESTIMATED ERROR:

Nothing is specified. A precision of ±0.1 pK units can be assumed (compiler).

REFERENCES:

COMPONENTS:

(1) Cesium tetraphenylborate (1-); $CsC_{24}H_{20}B$; [3087-82-9]

(2) Methanol; CH_4O; [67-56-1]

ORIGINAL MEASUREMENTS:

Alexander, R.; Parker, A. J.
J. Am. Chem. Soc. 1967, *89*, 5549-51.

VARIABLES:

One temperature: 25°C

PREPARED BY:

Orest Popovych

EXPERIMENTAL VALUES:

The formal (concentration) solubility product of $CsBPh_4$ in methanol was reported as;

$$pK_{s0} = 6.1 \ (K_{s0} \text{ units are mol}^2 \text{ dm}^{-6}).$$

The solubility can be estimated as $(9 \pm 1) \times 10^{-4}$ mol dm^{-3} (compiler).

AUXILIARY INFORMATION

METHOD/APPARATUS/PROCEDURE:

UV spectrophotometry on solutions saturated under nitrogen, or titration for the BPh_4^- anion with silver nitrate. No other details.

SOURCE AND PURITY OF MATERIALS:

Not stated.

ESTIMATED ERROR:

Nothing is specified.
A precision of ±0.1 pK units can be assumed (compiler).

REFERENCES:

COMPONENTS:

(1) Ammonium tetraphenylborate (1-); $C_{24}H_{20}BN$; [14637-34-4]

(2) Water; H_2O; [7732-18-5]

EVALUATOR:

Orest Popovych, Department of Chemistry, City University of New York, Brooklyn College, Brooklyn, N. Y. 11210, U. S. A.
November 1979

CRITICAL EVALUATION:

The solubility of ammonium tetraphenylborate (NH_4BPh_4) in aqueous solutions was published in four articles (1-4). In the two studies by Pflaum and Howick (1,2), the solubility was determined at 298 K by uv-spectrophotometry, but the two reported results differed drastically, being 1.07×10^{-4} mol dm^{-3} and 2.88×10^{-4} mol dm^{-3}, respectively. Such huge discrepancy cannot be rationalized on the basis of any of the obvious short-comings in the reported work. Thus, while it is true that the method and the precision of the temperature control were not specified in either of the articles, we can readily see from Siska's (3) data that the solubility of NH_4BPh_4 in aqueous solutions varies on the average by about 7×10^{-6} mol dm^{-3} per degree in the 293-303 K range. It is also true that in their first study Pflaum and Howick (1) used absorption coefficients ε_{max} for the BPh_4^- ion which were characteristic of acetonitrile solutions. Although nothing was specified with respect to the absorption coefficients in their subsequent article (2), it is probable that the ε_{max} values characteristic of acetonitrile solutions were used to calculate solubilities in aqueous solutions from absorption data throughout their work. The molar ε_{max} values used by Pflaum and Howick (1) were 3.225×10^3 and 2.110×10^3 at 266 nm and 274 nm, respectively, while the corresponding values reported for aqueous solutions are 3.25×10^3 and 2.06×10^3, respectively (5). (All molar absorption coefficients are in the units of dm^3 (cm mol)$^{-1}$). Thus, if Pflaum and Howick averaged the solubility values determined at the two wavelengths, they benefitted from a compensation of errors, which in the case of the solubility of $KBPh_4$ in water (1) led to a result in excellent agreement with other literature data (see critical evaluation for $KBPh_4$ in water). Certainly, the discrepancies between aqueous and acetonitrile ε_{max} values, which are of the order of 1-3%, could not account for the unreasonably large difference between the two solubility values reported in the two studies by Pflaum and Howick (1,2).

The third source of information on the solubility of NH_4BPh_4 in aqueous solutions -- the article by Siska (3) -- can offer only an indirect check on the validity of the results from the previous two studies. This is so because at zero ionic strength Siska reported a solubility only at 293 K, which was 2.52×10^{-4} mol dm^{-3}. An estimate of the solubility at 298 K can be made from Siska's data on the variation of the solubility as a function of the temperature, but at an ionic strength of 0.1 mol dm^{-3} maintained by sodium sulfate. If we use a linear interpolation in the function of log C vs. T^{-1}, where C is the solubility, the interpolated solubility value at 298.15 K turns out to be 3.27×10^{-4} mol dm^{-3} (at the ionic strength of 0.1 mol dm^{-3}). If the Davies equation is used to estimate the activity coefficient: $-\log y_\pm = [0.509\, I^{1/2}/(1 + I^{1/2})] + 0.1\, I$, where $\underline{I}$ is the ionic strength, we obtain for a 0.1 mol dm^{-3} solution the value $y_\pm = 0.771$, from which the solubility at zero ionic strength is estimated to be 2.52×10^{-4} mol dm^{-3}.

McClure and Rechnitz (4) measured the solubility of NH_4BPh_4 in a 0.1 mol dm^{-3} tris(hydroxymethyl)aminomethane buffer solution at 298.0 K and reported it as 3.4×10^{-4} mol dm^{-3}. This is in fair agreement with the result interpolated above from Siska's data, but a precise comparison and a calculation of the solubility at zero ionic strength from the data of McClure and Rechnitz is impossible, because the ionic strength of their buffer solution is not known exactly. Unfortunately, Siska's solubility values are very likely to be too low due to undersaturation (the suspensions were agitated for only 3 hours). They were definitely too low in the case of $KBPh_4$ in water. Nevertheless, the data of Siska as well as of McClure and Rechnitz suggest that it must be the second reported solubility value

COMPONENTS:	EVALUATOR:
(1) Ammonium tetraphenylborate (1-); $C_{24}H_{20}BN$; [14637-34-3] (2) Water; H_2O; [7732-18-5]	Orest Popovych, Department of Chemistry, City University of New York, Brooklyn College, Brooklyn, N. Y. 11210, U. S. A. November 1979

CRITICAL EVALUATION: (continued)

by Howick and Pflaum (2), i.e., 2.88 x 10^{-4} mol dm^{-3}, that we should place our reliance on at 298 K. Of course, the above value should be regarded as tentative at best. At other temperature, the only available solubility values are those reported by Siska (3), which are probably too low due to undersaturation.

REFERENCES:

1. Pflaum, R. T.; Howick, L. C. *Anal. Chem.* 1956, *28*, 1542.
2. Howick, L. C.; Pflaum, R. T. *Anal. Chim. Acta* 1958, *19*, 343.
3. Siska, E. *Magy. Kem. Foly.* 1976, *82*, 275.
4. McClure, J. E.; Rechnitz, G. A. *Anal. Chem.* 1966, *38*, 136.
5. Popovych, O.; Friedman, R. M. *J. Phys. Chem.* 1966, *70*, 1671.

COMPONENTS:

(1) Ammonium tetraphenylborate (1-); $C_{24}H_{24}BN$; [14637-34-4]

(2) Water; H_2O; [7732-18-5]

ORIGINAL MEASUREMENTS:

Pflaum, R. T.; Howick, L. C. *Anal. Chem.* 1956, *28*, 1542-4.

VARIABLES:

One temperature: 25°C

PREPARED BY:

Orest Popovych

EXPERIMENTAL VALUES:

The solubility of NH_4BPh_4 in water was reported as:

$$1.07 \times 10^{-4} \text{ mol dm}^{-3}.$$

The authors also reported the molar absorption coefficients for the BPh_4^- ion in acetonitrile solutions to be 3.225×10^3 and 2.110×10^3 dm^3 $(cm\ mol)^{-1}$ at 266 and 274 nm, respectively.

AUXILIARY INFORMATION

METHOD/APPARATUS/PROCEDURE:

Ultraviolet spectrophotometry on a Cary Model 11 recording spectrophotometer. Saturated solutions were prepared in conductivity water by an unspecified procedure. Method of controlling the temperature was not stated. The concentration of BPh_4^- in saturated solutions was obtained from spectrophotometric measurements at 266 and 274 nm by applying the molar absorption coefficients specified above.

SOURCE AND PURITY OF MATERIALS:

$NaBPh_4$ (J. T. Baker Chemical Co.) was used as received for pptns, but was recrystallized from acetone-hexane mixt for detn of absorption coefficients. NH_4BPh_4 was prepared by metathesis of NH_4Cl and $NaBPh_4$ and recrystallized from an acetonitrile-water mixture.

ESTIMATED ERROR:

Nothing is specified, but the precision is likely to be ±1% (compiler).

REFERENCES:

COMPONENTS:

(1) Ammonium tetraphenylborate (1-); $C_{24}H_{24}BN$; [14637-34-4]

(2) Water, H_2O; [7732-18-5]

ORIGINAL MEASUREMENTS:

Howick, L. C.; Pflaum, R. T.
Anal. Chim. Acta 1958, *19*, 343-7.

VARIABLES:

One temperature: 25°C

PREPARED BY:

Orest Popovych

EXPERIMENTAL VALUES:

The solubility of NH_4BPh_4 in water was reported to be 2.88×10^{-4} mol dm^{-3}.

AUXILIARY INFORMATION

METHOD/APPARATUS/PROCEDURE:

Saturated solutions were prepared both by agitating the suspensions at 25°C continuously and by agitating them first for a 0.5 hr at 40-50°C and then cooling to 25°C. When equilibrium was attained, the filtered solutions were analyzed for the BPh_4^- anion by UV spectrophotometry, using a Cary Model 11 recording spectrophotometer. The method of temperature control was not stated.

SOURCE AND PURITY OF MATERIALS:

$NaBPh_4$ (J. T. Baker Chemical Co.) was used as received. Other chemicals were of reagent grade. Deionized water was used. NH_4BPh_4 was prepared by reacting a 5% excess of freshly prepared $NaBPh_4$ solution with NH_4Cl. The product was recrystallized from acetone-water and analyzed for purity both by UV-spectrophotometry in acetonitrile and by titration with $HClO_4$ in HAc glacial to the crystal violet end point in anhydrous acetone (1).

ESTIMATED ERROR:

Nothing specified.

REFERENCES:

(1) Flaschka, H. *Chemist Analyst* 1955, *44*, 60.

COMPONENTS:

(1) Ammonium tetraphenylborate (1-); $C_{24}H_{24}BN$; [14637-34-4]
(2) Sodium sulfate; Na_2SO_4; [7757-82-6]
(3) Water; H_2O; [7732-18-5]

ORIGINAL MEASUREMENTS:

Siska, E. *Magy. Kem. Foly.* 1976, *82*, 275-8.

VARIABLES:
Temperature range: 3-50°C
Concentration of Na_2SO_4
pH

PREPARED BY:
Orest Popovych

EXPERIMENTAL VALUES:

In distilled water at 20°C, the solubility of NH_4BPh_4 was reported to be:

$$C = 2.39 \times 10^{-4} \text{ mol dm}^{-3}$$

and the corresponding solubility product, $K_{s0} = C^2$, as 5.71×10^{-8} mol^2 dm^{-6}.

With ionic strength varied by means of Na_2SO_4, the following solubilities, C, were reported for NH_4BPh_4 at 20°C in water:

Ionic strength/mol dm^{-3}	10^4C/mol dm^{-3}
0	2.52
0.05	2.84
0.1	2.84
0.3	2.80
0.5	2.52
0.7	2.08
1.0	1.92
2.0	0.88

Keeping the ionic strength constant at 0.1 mol dm^{-3} with sodium sulfate, the following solubilities C were obtained as a function of the temperature:

Continued...

AUXILIARY INFORMATION

METHOD/APPARATUS/PROCEDURE:

Amperometric titration. For details see compilation for $KBPh_4$ in water based on the same reference.

SOURCE AND PURITY OF MATERIALS:

Not specified. NH_4BPh_4 was prepared from the chloride by metathesis with $NaBPh_4$.

ESTIMATED ERROR:

Precision in solubility determination is ±2%.
Temperature control: ±1°C.

REFERENCES:

COMPONENTS:

(1) Ammonium tetraphenylborate (1-); $C_{24}H_{24}BN$; [14637-34-4]
(2) Sodium sulfate; Na_2SO_4; [7757-82-6]
(3) Water; H_2O; [7732-18-5]

ORIGINAL MEASUREMENTS:

Siska, E. *Magy. Kem. Foly.* 1976, *82*, 275-8.

VARIABLES:

PREPARED BY:

COMMENTS AND/OR ADDITIONAL DATA:

t/°C	$10^4C/mol\ dm^{-3}$
3	1.96
10	2.40
20	2.76
30	3.44
40	4.32
50	6.00

Keeping the ionic strength constant at 0.1 mol dm^{-3} with sodium sulfate and the temperature at 20 ± 1°C, the following solubilities C were obtained as a function of pH varied by means of acetic acid and sodium hydroxide:

pH	$10^4C/mol\ dm^{-3}$
2.7	2.57
4.0	2.67
4.3	2.75
4.4	2.60
4.7	2.74
4.8	2.64
5.7	2.61
6.5	2.74

The authors summarize their findings by stating that at an ionic strength of 0.1 mol dm^{-3}, a pH range of 2.7-6.5 and a temperature of 20 ± 1°C, the solubility of NH_4BPh_4 in aqueous solution is $(2.67 \pm 0.067) \times 10^{-4}$ mol dm^{-3}. The error apparently refers to precision.

METHOD/APPARATUS/PROCEDURE:

SOURCE AND PURITY OF MATERIALS:

ESTIMATED ERROR:

REFERENCES:

COMPONENTS:	ORIGINAL MEASUREMENTS:
(1) Ammonium tetraphenylborate (1-); $NH_4C_{24}H_{20}B$; [14637-34-4] (2) Tris(hydroxymethyl)aminoethane; $C_4H_{11}NO_3$; [77-86-1] (3) Acetic acid; $C_2H_4O_2$; [64-19-7] (4) Water; H_2O; [7732-18-5]	McClure, J. E.; Rechnitz, G. A. *Anal. Chem.* 1966, *38*, 136-9.
VARIABLES:	PREPARED BY:
One temperature: 24.8°C	Orest Popovych

EXPERIMENTAL VALUES:

The solubility of ammonium tetraphenylborate (NH_4BPh_4) in an aqueous solution of tris(hydroxymethyl)aminoethane (THAM) buffer at pH 5.1 was reported as:

$$3.4 \times 10^{-4} \text{ mol dm}^{-3}.$$

AUXILIARY INFORMATION

METHOD/APPARATUS/PROCEDURE:

UV- spectrophotometry according to the procedure of Howick and Pflaum (1). No other details.

SOURCE AND PURITY OF MATERIALS:

Baker reagent-grade NH_4Cl was the starting material. The buffer solution was prepared to contain 0.1 mol dm^{-3} THAM and 0.01 mol dm^{-3} acetic acid, adjusted to pH 5.1 with G. F. Smith reagent-grade $HClO_4$. The source of BPh_4^- was a solution of $Ca(BPh_4)_2$ in THAM prepared from Fisher Scientific reagent-grade $NaBPh_4$ by the procedure of Rechnitz et al. (2) and standardized by potentiometric titrn with KCl and RbCl.

ESTIMATED ERROR:

Not stated.
Temperature: ±0.3° C

REFERENCES:

1. Howick, L. C.; Pflaum, R. T. *Anal. Chim. Acta* 1958, *19*, 342.

2. Rechnitz, G. A.; Katz, S. A.; Zamochnick, S. B. *Anal. Chem.* 1963, *35*, 1322.

COMPONENTS:

(1) Ammonium tetraphenylborate (1-); $C_{24}H_{24}BN$; [14637-34-4]

(2) 2-Propanone (acetone); C_3H_6O; [67-64-1]

(3) Water; H_2O; [7732-18-5]

ORIGINAL MEASUREMENTS:

Kirgintsev, A. N.; Kozitskii, V. P. *Izvest. Akad. Nauk SSSR, Khim. Ser.* 1968, 1170-2.

VARIABLES:

Acetone-water composition
One temperature: 25.00°C

PREPARED BY:

Orest Popovych

EXPERIMENTAL VALUES:

The authors reported mass % of NH_4BPh_4 in the saturated solutions, defined as grams of the salt of 100 cm^3 of the solution. The solubilities have been recalculated to mol dm^{-3} by the compiler.

Vol. % water in acetone*	Solubility of NH_4BPh_4 (mass/vol.)%	$C/mol\ dm^{-3}$
0.007	5.60	0.166
2	7.56	0.224
4	8.87	0.263
8	10.00	0.297
12	9.91	0.294
15	9.51	0.282
20	8.70	0.258
25	7.26	0.215
30	5.88	0.174
37	4.36	0.129
45	2.70	0.0801
52	1.44	0.0427
60	0.605	0.0179
70	0.153	4.54×10^{-3}
80	0.044	1.30×10^{-3}

*Determined by weighing. Solvent volume was taken as the sum of the volumes of acetone and water, neglecting the effect of mixing.

AUXILIARY INFORMATION

METHOD/APPARATUS/PROCEDURE:

Evaporation and weighing. Saturated solutions were prepared by shaking the suspensions in a constant-temperature bath for 6 hours. Aliquots were removed through cotton plugs and weighed. Solvent was removed by evaporation in a stream of air, followed by desiccation under P_2O_5 and under vacuum. The solid phase contained no solvent when recrystallized from acetone or acetone-water mixtures.

SOURCE AND PURITY OF MATERIALS:

See compilation sheet for $KBPh_4$ in acetone-water, based on the above reference. NH_4BPh_4 was prepared and purified as $KBPh_4$ by metathesis of $NaBPh_4$ and NH_4Cl.

ESTIMATED ERROR:

Precision ±0.5%
Temperature control: ±0.05°C

REFERENCES:

COMPONENTS:

(1) Ammonium tetraphenylborate (1-); $C_{24}H_{24}BN$; [14637-34-4]

(2) 1-Methyl-2-pyrrolidinone (N-Methyl-2-pyrrolidone); C_5H_9NO; [872-50-4]

ORIGINAL MEASUREMENTS:

Virtanen, P. O. I.; Kerkelä, R. *Suom. Kemistil.* 1969, *B42*, 29-33.

VARIABLES:

Two temperatures: 25.00 °C and 45.00 °C.

PREPARED BY:

Orest Popovych

EXPERIMENTAL VALUES:

The solubility of NH_4BPh_4 in N-methyl-2-pyrrolidone was reported to be:

1.21 mol dm^{-3} at 25 °C and 1.24 mol dm^{-3} at 45 °C.

The corresponding solubility product at 25 °C, calculated as the square of the solubility, was reported in the form $pK_{s0} = -0.16$, where K_{s0} units are $mol^2\ dm^{-6}$. The pK_{s0} value at 45 °C was not reported.

AUXILIARY INFORMATION

METHOD/APPARATUS/PROCEDURE:

The suspensions were shaken in thermostatted water-jacketed flasks for 1 day at 50 °C, followed by 1 day at 25 °C or 45 °C, respectively. Saturated solutions were analyzed by precipitating the NH_4BPh_4 from aliquots in aqueous solution.

SOURCE AND PURITY OF MATERIALS:

N-Methyl-2-pyrrolidone (General Aniline and Film Co.) was purified as in the literature (1). NH_4BPh_4 was prepared by metathesis of NH_4Cl and $NaBPh_4$ in water, followed by double recrystallization from an acetone-water mixture and drying *in vacuo*.

ESTIMATED ERROR:

Not specified.
Temperature control: ±0.02 °C

REFERENCES:

(1) Virtanen, P. O. I. *Suom. Kemistil.* 1966, *B39*, 257.

COMPONENTS:

(1) N,N'- bis(3-aminopropyl)-1,4-butanediamine (spermine) tetrakis-tetraphenylborate (1-); $C_{106}H_{110}B_4N_4$;

(2) Water; H_2O; [7732-18-5]

ORIGINAL MEASUREMENTS:

Zeidler, L. *Hoppe-Seyler's Z. Physiol. Chem.* 1952, *291*, 177-8.

VARIABLES:

Presumably room temperature

PREPARED BY:

Orest Popovych

EXPERIMENTAL VALUES:

The solubility of spermine tetrakis-tetraphenylborate was reported as 0.02%, probably meaning 0.02 g in 100 cm^3 of saturated solution. If this interpretation is correct, the solubility corresponds to $1._3 \times 10^{-4}$ mol dm^{-3} (compiler).

AUXILIARY INFORMATION

METHOD/APPARATUS/PROCEDURE:

Nothing specified.

SOURCE AND PURITY OF MATERIALS:

The salt was prepared by reacting the amine in neutral or weakly acidic solution with a freshly prepared solution of $NaBPh_4$ ("Kalignost" from Heyl & Co.). Analysis of the product yielded 3.69% N, as compared to 3.78% theoretical.

ESTIMATED ERROR:

Nothing specified.

REFERENCES:

COMPONENTS:	ORIGINAL MEASUREMENTS:
(1) 1,4-Butanediamine (putrescine) bis-tetraphenylborate (1-); $C_{52}H_{54}B_2N_2$; (2) Water; H_2O; [7732-18-5]	Zeidler, L. *Hoppe-Seyler's Z. Physiol. Chem.* 1952, *291*, 177-8.
VARIABLES:	PREPARED BY:
Presumably room temperature	Orest Popovych

EXPERIMENTAL VALUES:

The solubility of putrescine bis-tetraphenylborate was reported as 0.027%, probably meaning 0.027 g in 100 cm^3 of saturated solution. If this interpretation is correct, the solubility corresponds to 3.7×10^{-4} mol dm^{-3} (compiler).

AUXILIARY INFORMATION

METHOD/APPARATUS/PROCEDURE:	SOURCE AND PURITY OF MATERIALS:
Nothing specified.	The salt was prepared by reacting the amine in neutral or weakly acidic solution with a freshly prepared solution of $NaBPh_4$ ("Kalignost" from Heyl & Co.). Analysis of the product yielded 3.62% N, as compared to 3.85% theoretical.
	ESTIMATED ERROR:
	Nothing specified.
	REFERENCES:

COMPONENTS:

(1) Butylammonium tetraphenylborate (1-); $C_{28}H_{32}BN$; [69502-97-2]

(2) Water; H_2O; [7732-18-5]

ORIGINAL MEASUREMENTS:

Howick, L. C.; Pflaum, R. T. *Anal. Chim. Acta* 1958, *19*, 343-7.

VARIABLES:

One temperature: 25°C

PREPARED BY:

Orest Popovych

EXPERIMENTAL VALUES:

The solubility of butylammonium tetraphenylborate in water was reported as:

$$1.12 \times 10^{-3} \text{ mol dm}^{-3}.$$

For a critical evaluation of the data from this study, see the evaluation for NH_4BPh_4 in water.

AUXILIARY INFORMATION

METHOD/APPARATUS/PROCEDURE:

Saturated solutions were prepared both by agitating the suspensions at 25°C continuously and by agitating them first for a 0.5 hr at 40-50°C and then cooling to 25°C. When equilibrium was attained, the filtered solutions were analyzed for the BPh_4^- anion by UV spectrophotometry, using a Cary Model 11 recording spectrophotometer. The method of temperature control was not stated.

SOURCE AND PURITY OF MATERIALS:

See the compilation for NH_4BPh_4 in water based on the same reference. The amine hydrochloride from which the tetraphenylborate was prepared was an Eastman White Label product.

ESTIMATED ERROR:

Nothing specified.

REFERENCES:

COMPONENTS:

(1) Butyltriisopentylammonium tetraphenylborate (1-); (Triisoamyl-n-butylammonium tetraphenylborate (1-); $C_{33}H_{62}BN$; [16742-92-0]

(2) Sodium hydroxide; NaOH: [1310-73-2]

(3) Water; H_2O; [7732-18-5]

ORIGINAL MEASUREMENTS:

Popovych, O.; Friedman, R. M. *J. Phys. Chem.* 1966, *70*, 1671-3.

VARIABLES:

One temperature: 25.00°C

PREPARED BY:

Orest Popovych

EXPERIMENTAL VALUES:

The authors report the solubility of triisoamyl-n-butylammonium tetraphenylborate (TAB BPh_4) in water determined directly by uv-spectrophotometry in the presence of 10^{-5} mol dm^{-3} NaOH as an uncertain value:

$$C = 1.4 \times 10^{-7} ? \text{mol dm}^{-3}.$$

Because of the tendency of TAB BPh_4 to decompose upon prolonged equilibration with water, the authors also calculated the solubility indirectly from the transfer activity coefficients (medium effects) as follows:

$$\log {}_{m}y_{\pm}^{2}(\text{TAB BPh}_4) = \log {}_{m}y_{\pm}^{2}(\text{TAB Pi}) + \log {}_{m}y_{\pm}^{2}(\text{KBPh}_4) - \log {}_{m}y_{\pm}^{2}(\text{KPi}) \quad (1)$$

and

$$\log {}_{m}y_{\pm}^{2} = \log \frac{K_{s0}^{\circ}\ (\text{water})}{K_{s0}^{\circ}\ (\text{methanol})} \quad (2)$$

Substituting on the rhs of Equation (1) values of ${}_{m}y_{\pm}^{2}$ obtained from experimental determinations of solubilities and Equation (2), the authors obtained:

$$\log {}_{m}y_{\pm}^{2}(\text{TAB BPh}_4) = -5.398 + (-2.300) - 1.101 = -8.799$$

Using the above result as well as K_{s0}° (methanol) = 7.36×10^{-6} in Equ. (2), they obtained K_{s0}° (water) = 1.17×10^{-14} mol^2 dm^{-6} (units by compiler), from which the solubility C = 1.08×10^{-7} mol dm^{-3}.

In the above equations, the transfer activity coefficients refer to the transfer from water to methanol. Pi = picrate ion.

AUXILIARY INFORMATION

METHOD/APPARATUS/PROCEDURE:

Ultraviolet spectrophotometry using a Cary Model 14 spectrophotometer. Saturation achieved by shaking the salt suspensions for 2 weeks in water-jacketed flasks. Solutions filtered and analyzed at 266 and 274 nm. All solutions and containers were deaerated.

SOURCE AND PURITY OF MATERIALS:

TAB BPh_4 was synthesized and purified by the method of Coplan and Fuoss (1).

ESTIMATED ERROR:

Not specified.
Temperature: ±0.01°C.

REFERENCES:

(1) Coplan, M. A.; Fuoss, R. M. *J. Phys. Chem.* 1964, *68*, 1177.

COMPONENTS:

(1) Butyltriisopentylammonium tetraphenylborate (1-) (Triisoamyl-n-butylammonium tetraphenylborate); $C_{33}H_{62}BN$; [16742-92-0]
(2) Lithium chloride; LiCl; [7447-41-8]
(3) Ethanol; C_2H_6O; [64-17-5]
(4) Water; H_2O; [7732-18-5]

ORIGINAL MEASUREMENTS:

Dill, A. J.; Popovych, O.
J. Chem. Eng. Data 1969, *14*, 240-3.

VARIABLES: Ethanol-water composition. LiCl concentration varied from 0 to 200 times molar solubility of $C_{33}H_{62}BN$. One temperature: 25.00°C

PREPARED BY:

Orest Popovych

EXPERIMENTAL VALUES:

The authors report the solubility of triisoamyl-n-butylammonium tetraphenylborate (TAB BPh_4) in ethanol-water mixtures without added LiCl. The solubilities with added LiCl are not reported, but they were used to calculate the activity coefficients of TAB BPh_4 by a procedure identical with that described in the compilation for $KBPh_4$ in ethanol-water mixtures.

Mass % ethanol in water	Solubility of TAB BPh_4, 10^4C/mol dm^{-3}
100.0	11.8
90.0	7.30
80.0	5.94
78.1	4.50
70.0	2.97
60.6	1.48
60.0	1.21
50.0*	0.60
46.0	0.512
40.0	0.210

*Graphically interpolated. Activity coefficient data and solubility products are tabulated below. The units of K°_{s0} are mol^2 dm^{-6} (compiler)

Mass % ethanol in water	$\alpha_o(1)$	$y_{\pm,o}$	A_1	A_2	A_3	$K^\circ_{s0} = (C_o \alpha_o y_\pm)^2$
100.0	0.886	0.842	2.44	-5.86	8.38	7.74×10^{-7}
78.1	0.986	0.946	0.703	---	---	1.76×10^{-7}
60.6	0.998	0.957	1.17	-0.985	----	2.00×10^{-8}

(The above symbols are from equations in the compilation for $KBPh_4$ in methanol.)

AUXILIARY INFORMATION

METHOD/APPARATUS/PROCEDURE:

Ultraviolet spectrophotometry using a Cary Model 14 spectrophotometer. A solution was considered saturated when successive weekly analyses agreed to about 1%. This required 2 weeks of equilibration for solutions without added LiCl and one month for solutions with added LiCl. Solutions filtered and analyzed at 266 and 274 nm. All solutions and containers were deaerated.

SOURCE AND PURITY OF MATERIALS:

TAB BPh_4 was synthesized and purified by the method of Coplan and Fuoss (2). Purification of ethanol and the preparation of ethanol-water mixtures is described on the compilation sheet for $KBPh_4$ in ethanol-water mixtures.

ESTIMATED ERROR: (For the solubility)

Precision ±1%
Accuracy: ±3% (authors)
Temperature: ±0.01°C.

REFERENCES:

(1) Dill, A. J.; Popovych, O. *J. Chem. Eng. Data* 1969, *14*, 156.
(2) Coplan, M. A.; Fuoss, R. M. *J. Phys. Chem.* 1964, *68*, 1177.

COMPONENTS:
(1) Butyltriisopentylammonium tetraphenylborate (1-) (Triisoamyl-n-butylammonium tetraphenylborate (1-); $C_{33}H_{62}BN$; [16742-92-0]
(2) Toluene; C_7H_8; [108-88-3]
(3) Isopropyl alcohol; C_3H_8O; [67-63-0]
(4) Water; H_2O; [7732-18-5]

ORIGINAL MEASUREMENTS:
Popovych, O. *Anal. Chem.* 1966, *38*, 558-63.

VARIABLES:
One temperature: 25.00°C

PREPARED BY:
Orest Popovych

EXPERIMENTAL VALUES:

The author reports as the solubility of triisoamyl-n-butylammonium tetraphenylborate (TAB BPh_4) in the toluene-isopropyl-alcohol-water mixture known as the ASTM medium:* $C = 1.09 \times 10^{-4}$ mol dm^{-3}.

From a $K_{diss} = 5.62 \times 10^{-6}$ mol dm^{-3}, the degree of dissociation α in saturated solution was calculated to be 0.238 and the mean ionic activity coefficient (volume basis) was estimated from the limiting Debye-Hückel law as $y_{\pm}^2 = 0.691$. Combining these, the author arrived at the solubility product of TAB BPh_4 in the ASTM solvent:

$$K_{s0}^{\circ} = 4.67 \times 10^{-10} \text{ mol}^2 \text{ dm}^{-6},$$

(where $K_{s0}^{\circ} = (C \alpha y_{\pm})^2$).

*American Society for Testing Materials specifies this solvent for acid-base measurements on petroleum products.

AUXILIARY INFORMATION

METHOD/APPARATUS/PROCEDURE:
Electrolytic conductance of diluted saturated solutions, using a calibration curve for TAB BPh_4 concentration. The conductance apparatus was a Wayne-Kerr Universal Bridge B221 with a platinum cell. Saturation was achieved by shaking for at least 2 weeks on a Burrell wrist-action shaker in water-jacketed flasks.

SOURCE AND PURITY OF MATERIALS:
TAB BPh_4 was synthesized and purified essentially by the method of Coplan and Fuoss (1). For the purification of the solvents, the authors referred to a literature source (2).

ESTIMATED ERROR:
Not specified.
Temperature: ±0.01°C.

REFERENCES:
(1) Coplan, M. A.; Fuoss, R. M. *J. Phys. Chem.* 1964, *68*, 1177.
(2) Popovych, O. *J. Phys. Chem.* 1962, *66*, 915.

COMPONENTS:

(1) Butyltriisopentylammonium tetraphenylborate (1-); (triisoamyl-n-butylammonium tetraphenylborate (1-); $C_{33}H_{62}BN$; [16742-92-0]

(2) Lithium chloride; LiCl; [7447-41-8]

(3) Ethanol; C_2H_6O; [64-17-5]

ORIGINAL MEASUREMENTS:

Dill, A. J.; Popovych, O.
J. Chem. Eng. Data 1969, *14*, 240-3.

VARIABLES:

LiCl concentration varied from 0 to 200 times the solubility of $C_{33}H_{62}BN$. One temperature: 25.00°C

PREPARED BY:

Orest Popovych

EXPERIMENTAL VALUES:

The solubility of triisoamyl-n-butylammonium tetraphenylborate in ethanol without added LiCl was reported as 1.18×10^{-3} mol dm^{-3}.

Solubilities with added LiCl were not reported as such, only the activity coefficient derived from the variation of the solubility as a function of ionic strength varied by means of LiCl (for details see compilation for $KBPh_4$ in ethanol-water mixtures based on the same literature reference). Using an association constant K_A = 192 mol^{-1} dm^3 (1) the authors computed the degree of dissociation in saturated solution to be $\alpha_\circ = 0.886$. The experimentally determined mean ionic activity coefficient in saturated solution was reported to be $y_{\pm,\circ} = 0.842$. Combining these:

$$K^\circ_{s0} = (1.18 \times 10^{-3} \text{ mol dm}^{-3} \times 0.886 \times 0.842)^2 = 7.74 \times 10^{-7} \text{ mol}^2 \text{ dm}^{-6}.$$

AUXILIARY INFORMATION

METHOD/APPARATUS/PROCEDURE:

Ultraviolet spectrophotometry using a Cary Model 14 spectrophotometer. A solution was considered saturated when successive weekly analyses agreed to about 1%. This required 2 weeks of equilibration for solutions without added LiCl and one month for solutions with added LiCl. Solutions filtered and analyzed at 266 and 274 nm. All solutions and containers were deaerated.

SOURCE AND PURITY OF MATERIALS:

Triisoamyl-n-butylammonium tetraphenylborate was synthesized and purified by the method of Coplan and Fuoss (2). Purification of ethanol was described in the compilation for $KBPh_4$ in ethanol-water system based on the same reference.

ESTIMATED ERROR: (For the solubility)

Precision ±1%
Accuracy ±3% (authors)
Temperature control: ±0.01°C

REFERENCES:

(1) Dill, A. J.; Popovych, O. *J. Chem. Eng. Data* 1969, *14*, 156.

(2) Coplan, M. A.; Fuoss, R. M. *J. Phys. Chem.* 1964, *68*, 1177.

COMPONENTS:

(1) Butyltriisopentylammonium tetraphenylborate (1-) (Triisoamyl-n-butylammonium tetraphenylborate (1-)); $C_{33}H_{62}BN$; [16742-92-0]

(2) Methanol; CH_4O; [67-56-1]

ORIGINAL MEASUREMENTS:

Popovych, O.; Friedman, R. M. *J. Phys. Chem.* 1966, *70*, 1671-3.

VARIABLES:

One temperature: 25.00°C

PREPARED BY:

Orest Popovych

EXPERIMENTAL VALUES:

The authors report the solubility of triisoamyl-n-butylammonium tetraphenylborate (TAB BPh_4) as:

$$C = 3.60 \times 10^{-3} \text{ mol dm}^{-3}.$$

The solubility (ion-activity) product, K°_{s0}, was calculated by the authors as $(C\alpha y_\pm)^2$, where the degree of dissociation α was calculated from the following relationship using a literature value (1) for the ion-pair dissociation constant K_A:

$$\alpha = \frac{-1 + (1 + 4K_A C y_\pm^2)^{1/2}}{2K_A C y_\pm^2} .$$

The activity coefficient $y_\pm$ was estimated from the Debye-Hückel equation in the form:

$$-\log y_\pm^2 = \frac{3.803\ (C\alpha)^{1/2}}{1 + 0.5099\ \mathring{a}(C\alpha)^{1/2}} \quad \text{using } \mathring{a} = 0.7 \text{ nm}.$$

The above calculations yielded $\alpha = 0.930$ and $y_\pm^2 = 0.657$ from which $K^\circ_{s0} = 7.36 \times 10^{-6}$ mol^2 dm^{-6} (compiler's units).
The molar absorption coefficients of the tetraphenylborate ion used to calculate the solubilities were reported to be 3.00×10^3 and 2.12×10^3 dm^3 $(cm\ mol)^{-1}$ at 266 and 274 nm, respectively.

AUXILIARY INFORMATION

METHOD/APPARATUS/PROCEDURE:

Ultraviolet spectrophotometry using a Cary Model 14 spectrophotometer. Saturation achieved by shaking the salt suspensions for 2 weeks in water-jacketed flasks. Solutions filtered and analyzed at 266 and 274 nm. All solutions and containers were deaerated.

SOURCE AND PURITY OF MATERIALS:

Source and purification of methanol the same as described in the compilation for potassium tetraphenylborate. TAB BPh_4 was synthesized and purified by the method of Coplan and Fuoss (1)

ESTIMATED ERROR:

None stated, but the precision is known to be about ±1% for the solubility. Temperature: ±0.01°C.

REFERENCES:

(1) Coplan, M. A.; Fuoss, R. M. *J. Phys. Chem.* 1964, *68*, 1177.

COMPONENTS:

(1) Dimethylammonium tetraphenylborate (1-); $C_{26}H_{28}BN$; [69502-98-3]

(2) Water; H_2O; [7732-18-5]

ORIGINAL MEASUREMENTS:

Howick, L. C.; Pflaum, R. T.
Anal. Chim. Acta 1958, *19*, 343-7.

VARIABLES:

One temperature: 25°C

PREPARED BY:

Orest Popovych

EXPERIMENTAL VALUES:

The solubility of dimethylammonium tetraphenylborate in water was reported as:

$$1.63 \times 10^{-3} \text{ mol dm}^{-3}.$$

For a critical evaluation of the data from this study, see the evaluation for NH_4BPh_4 in water.

AUXILIARY INFORMATION

METHOD/APPARATUS/PROCEDURE:

Saturated solutions were prepared both by agitating the suspensions at 25°C continuously and by agitating them first for a 0.5 hr at 40-50°C and then cooling to 25°C. When equilibrium was attained, the filtered solutions were analyzed for the BPh_4^- anion by UV spectrophotometry, using a Cary Model 11 recording spectrophotometer. The method of temperature control was not stated.

SOURCE AND PURITY OF MATERIALS:

See the compilation for NH_4BPh_4 in water based on the same reference. The amine hydrochloride used to prepare the tetraphenylborate was an Eastman White Label product.

ESTIMATED ERROR:

Nothing specified.

REFERENCES:

COMPONENTS:

(1) Ethylammonium tetraphenylborate (1-); $C_{26}H_{28}BN$; [53694-97-6]

(2) Water; H_2O; [7732-18-5]

ORIGINAL MEASUREMENTS:

Howick, L. C.; Pflaum, R. T. *Anal. Chim. Acta* 1958, *19*, 343-7.

VARIABLES:

One temperature: 25°C

PREPARED BY:

Orest Popovych

EXPERIMENTAL VALUES:

The solubility of ethylammonium tetraphenylborate in water was reported as:

$$2.83 \times 10^{-3} \text{ mol dm}^{-3}.$$

For a critical evaluation of the data from this study, see the evaluation for NH_4BPh_4 in water.

AUXILIARY INFORMATION

METHOD/APPARATUS/PROCEDURE:

Saturated solutions were prepared both by agitating the suspensions at 25°C continuously and by agitating them first for a 0.5 hr at 40-50°C and then cooling to 25°C. When equilibrium was attained, the filtered solutions were analyzed for the BPh_4^- anion by UV spectrophotometry, using a Cary Model 11 recording spectrophotometer. The method of temperature control was not stated.

SOURCE AND PURITY OF MATERIALS:

See the compilation for NH_4BPh_4 in water based on the same reference. The amine hydrochloride from which the tetraphenylborate was prepared was an East White Label product.

ESTIMATED ERROR:

Nothing specified.

REFERENCES:

COMPONENTS:

(1) Guanidine tetraphenylborate (1-); $C_{25}H_{26}BN_3$;

(2) Water; H_2O; [7732-18-5]

ORIGINAL MEASUREMENTS:

Zeidler, L. *Hoppe-Seyler's Z. Physiol. Chem.* 1952, *291*, 177-8.

VARIABLES:

Presumably room temperature

PREPARED BY:

Orest Popovych

EXPERIMENTAL VALUES:

The solubility of guanidine tetraphenylborate was reported as 0.14%, probably meaning 0.14 g in 100 cm^3 of saturated solution. If this interpretation is correct, the solubility corresponds to 3.7 x 10^{-3} mol dm^{-3} (compiler).

AUXILIARY INFORMATION

METHOD/APPARATUS/PROCEDURE:

Nothing specified.

SOURCE AND PURITY OF MATERIALS:

The salt was prepared by reacting the amine in neutral or weakly acidic solution with a freshly prepared solution of $NaBPh_4$ ("Kalignost" from Heyl & Co.). Analysis of the product gave 10.43% N as compared to 11.08% theoretical.

ESTIMATED ERROR:

Nothing specified.

REFERENCES:

COMPONENTS:

(1) Histamine bis-tetraphenylborate (1-); $C_{53}H_{51}B_2N_3$;

(2) Water; H_2O; [7732-18-5]

ORIGINAL MEASUREMENTS:

Zeidler, L. *Hoppe-Seyler's Z. Physiol. Chem.* 1952, *291*, 177-8.

VARIABLES:

Presumably room temperature

PREPARED BY:

Orest Popovych

EXPERIMENTAL VALUES:

The solubility of histamine bis-tetraphenylborate was reported as 0.01%, probably meaning 0.01 g in 100 cm^3 of saturated solution. If this interpretation is correct, the solubility corresponds to $1._3 \times 10^{-4}$ mol dm^{-3} (compiler).

AUXILIARY INFORMATION

METHOD/APPARATUS/PROCEDURE:

Nothing specified.

SOURCE AND PURITY OF MATERIALS:

The salt was prepared by reacting the amine in neutral or weakly acidic solution with a freshly prepared solution of $NaBPh_4$ ("Kalignost" from Heyl & Co.). Analysis of the product gave 5.50% N, as compared to 5.59% theoretical.

ESTIMATED ERROR:

Nothing specified.

REFERENCES:

COMPONENTS:

(1) 1H-Imidazole-4-ethanamine (histidine) tetraphenylborate (1-); $C_{30}H_{30}BN_3O_2$;

(2) Water; H_2O; [7732-18-5]

ORIGINAL MEASUREMENTS:

Zeidler, L. *Hoppe-Seyler's Z. Physiol. Chem.* 1952, *291*, 177-8.

VARIABLES:

Presumably room temperature

PREPARED BY:

Orest Popovych

EXPERIMENTAL VALUES:

The solubility of histidine tetraphenylborate was reported as 0.24%, probably meaning 0.24 g in 100 cm^3 of saturated solution. If this interpretation is correct, the solubility corresponds to 5.1×10^{-3} mol dm^{-3} (compiler).

AUXILIARY INFORMATION

METHOD/APPARATUS/PROCEDURE:

Nothing specified.

SOURCE AND PURITY OF MATERIALS:

The salt was prepared by reacting the amine in neutral or weakly acidic solution with a freshly prepared solution of $NaBPh_4$ ("Kalignost" from Heyl & Co.). Analysis of the product yielded 8.01% N, as compared to 8.64% theoretical.

ESTIMATED ERROR:

Nothing specified.

REFERENCES:

COMPONENTS:

(1) Methylammonium tetraphenylborate (1-); $C_{25}H_{26}BN$; [60337-02-2]

(2) Water; H_2O; [7732-18-5]

ORIGINAL MEASUREMENTS:

Howick, L. C.; Pflaum, R. T.
Anal. Chim. Acta 1958, *19*, 343-7.

VARIABLES:

One temperature: 25°C

PREPARED BY:

Orest Popovych

EXPERIMENTAL VALUES:

The solubility of methylammonium tetraphenylborate in water was reported as:

3.63×10^{-3} mol dm^{-3}.

For a critical evaluation of the data from this study, see the evaluation for NH_4BPh_4 in water.

AUXILIARY INFORMATION

METHOD/APPARATUS/PROCEDURE:

Saturated solutions were prepared both by agitating the suspensions at 25°C continuously and by agitating them first for a 0.5 hr at 40-50°C and then cooling to 25°C. When equilibrium was attained, the filtered solutions were analyzed for the BPh_4^- anion by UV spectrophotometry, using a Cary Model 11 recording spectrophotometer. The method of temperature control was not stated.

SOURCE AND PURITY OF MATERIALS:

See the compilation for NH_4BPh_4 in water based on the same reference. Methylammonium tetraphenylborate, prepared from the hydrochloride (Eastman White Label) was not recrystallized before use.

ESTIMATED ERROR:

Nothing specified.

REFERENCES:

COMPONENTS:

(1) 1,5-Pentanediamine (cadaverine) bis-tetraphenylborate (1-); $C_{53}H_{56}B_2N_2$;

(2) Water, H_2O; [7732-18-5]

ORIGINAL MEASUREMENTS:

Zeidler, L. *Hoppe-Seyler's Z. Physiol. Chem.* 1952, *291*, 177-8.

VARIABLES:

Presumably room temperature

PREPARED BY:

Orest Popovych

EXPERIMENTAL VALUES:

The solubility of cadaverine bis-tetraphenylborate was reported as 0.031%, probably meaning 0.031 g in 100 cm^3 of saturated solution. If this interpretation is correct, the solubility corresponds to 4.2×10^{-4} mol dm^{-3} (compiler).

AUXILIARY INFORMATION

METHOD/APPARATUS/PROCEDURE:

Nothing specified.

SOURCE AND PURITY OF MATERIALS:

The salt was prepared by reacting the amine in neutral or weakly acidic solution with freshly prepared solution of $NaBPh_4$ ("Kalignost" from Heyl & Co.). Analysis of the product gave 3.45% N as compared to 3.77% theoretical.

ESTIMATED ERROR:

Nothing specified.

REFERENCES:

COMPONENTS:

(1) Propylammonium tetraphenylborate (1-); $C_{27}H_{30}BN$; [6928-94-5]

(2) Water; H_2O; [7732-18-5]

ORIGINAL MEASUREMENTS:

Howick, L. C.; Pflaum, R. T. *Anal. Chim. Acta* 1958, *19*, 343-7.

VARIABLES:

One temperature: 25°C

PREPARED BY:

Orest Popovych

EXPERIMENTAL VALUES:

The solubility of propylammonium tetraphenylborate in water was reported as:

$$9.03 \times 10^{-4} \text{ mol dm}^{-3}.$$

For a critical evaluation of the data from this study, see the evaluation for NH_4BPh_4 in water.

AUXILIARY INFORMATION

METHOD/APPARATUS/PROCEDURE:

Saturated solutions were prepared both by agitating the suspensions at 25°C continuously and by agitating them first for a 0.5 hr at 40-50°C and then cooling to 25°C. When equilibrium was attained, the filtered solutions were analyzed for the BPh_4^- anion by UV spectrophotometry, using a Cary Model 11 recording spectrophotometer. The method of temperature control was not stated.

SOURCE AND PURITY OF MATERIALS:

See the compilation for NH_4BPh_4 in water based on the same reference. The amine hydrochloride from which the tetraphenylborate was prepared was an Eastman White Label product.

ESTIMATED ERROR:

Nothing specified.

REFERENCES:

COMPONENTS:

(1) Pyridinium tetraphenylborate (1-); $C_{29}H_{26}BN$; [50328-28-4]

(2) Water, H_2O; [7732-18-5]

ORIGINAL MEASUREMENTS:

Howick, L. C.; Pflaum, R. T.
Anal. Chim. Acta 1958, *19*, 343-7.

VARIABLES:

One temperature: 25°C

PREPARED BY:

Orest Popovych

EXPERIMENTAL VALUES:

The solubility of pyridinium tetraphenylborate in water was reported as:

$$1.99 \times 10^{-4} \text{ mol dm}^{-3}.$$

For a critical evaluation of the data from this study, see the evaluation for NH_4BPh_4 in water.

AUXILIARY INFORMATION

METHOD/APPARATUS/PROCEDURE:

Saturated solutions were prepared both by agitating the suspensions at 25°C continuously and by agitating them first for a 0.5 hr at 40-50°C and then cooling to 25°C. When equilibrium was attained, the filtered solutions were analyzed for the BPh_4^- anion by UV spectrophotometry, using a Cary Model 11 recording spectrophotometer. The method of temperature control was not stated.

SOURCE AND PURITY OF MATERIALS:

See the compilation for NH_4BPh_4 in water based on the same reference. To an ethanolic solution of pyridine (Eastman White Label) $HClO_4$ was added slowly and the perchlorate recrystallized from the water-ethanol was used to prepare the tetraphenylborate.

ESTIMATED ERROR:

Nothing specified.

REFERENCES:

COMPONENTS:

(1) Tetra-*n*-butylammonium tetraphenylborate (1-); $C_{40}H_{56}BN$; [15522-59-5]

(2) Sodium hydroxide; NaOH; [1310-73-2]

(3) Water; H_2O; [7732-18-5]

ORIGINAL MEASUREMENTS:

Popovych, O; Friedman, R. M. *J. Phys. Chem.* 1966, *70*, 1671-3.

VARIABLES:

One temperature: 25.00°C

PREPARED BY:

Orest Popovych

EXPERIMENTAL VALUES:

The authors report the solubility of tetra-n-butylammonium tetraphenylborate ($Bu_4N\ BPh_4$) in water determined directly by uv-spectrophotometry in the presence of 10^{-5} M NaOH as an uncertain value:

$$C = 3.4 \times 10^{-6}?\ \text{mol dm}^{-3}.$$

Because of the tendencey of $Bu_4N\ BPh_4$ to decompose upon prolong equilibration with water, the authors also calculated the solubility indirectly, via transfer activity coefficients (medium effects) as follows:

$$\log {}_m y_{\pm}^2(Bu_4N\ BPh_4) = \log {}_m y_{\pm}^2(Bu_4N\ Pi) + \log {}_m y_{\pm}^2(KBPh_4) - \log {}_m y_{\pm}^2(KPi) \quad (1)$$

where ${}_m y_{\pm}$ is the mean molar medium effect for the eletrolyte in methanol (i.e., the activity coefficient for the transfer water → methanol) and Pi is the picrate ion.

Values of $\log {}_m y_{\pm}$ for the electrolytes on the rhs of the above equation were obtained from the experimentally determined solubility products:

$$\log {}_m y_{\pm}^2 = \log \frac{K_{s0}^{\circ}\ (\text{water})}{K_{s0}^{\circ}\ (\text{methanol})} \quad (2)$$

then: $\log {}_m y_{\pm}^2(Bu_4N\ BPh_4) = -4.438 + (-2.300) - 1.101 = -7.839$

introducing in equation (2) the above result as well as $K_{s0}^{\circ} = 4.62 \times 10^{-6}$ $\text{mol}^2\ \text{dm}^{-6}$ in methanol, the authors obtained $K_{s0}^{\circ} = 6.69 \times 10^{-14}$ (water) from which the solubility $C = 2.59 \times 10^{-7}$ mol dm^{-3}.

AUXILIARY INFORMATION

METHOD/APPARATUS/PROCEDURE:

UV spectrophotometry using a Cary Model 14 spectrophotometer. Saturation achieved by shaking the salt suspensions for 2 weeks in water-jacketed flasks. Solutions filtered and analyzed at 266 and 274 nm. All solutions and containers were deaerated.

SOURCE AND PURITY OF MATERIALS:

$Bu_4N\ BPh_4$ was synthesized and purified as described in the literature (1).

ESTIMATED ERROR:

Not specified.
Temperature: ±0.01°C

REFERENCES:

(1) Accascina, F.; Petrucci, S. Fuoss, R. M. *J. Am. Chem. Soc.* 1959, *81*, 1301.

COMPONENTS:

(1) Tetra-n-butylammonium tetraphenylborate (1-); $C_{40}H_{56}BN$; [15522-59-5]

(2) 1,1-Dichloroethane; $C_2H_4Cl_2$; [75-34-3]

ORIGINAL MEASUREMENTS:

Abraham, M. H.; Danil de Namor, A. F. *J. Chem. Soc. Faraday Trans. 1* 1976, *72*, 955-62.

VARIABLES:

One temperature: 25°C

PREPARED BY:

Orest Popovych

EXPERIMENTAL VALUES:

The authors reported the solubility of Bu_4NBPh_4 in 1,1-dichloroethane as:
5.09×10^{-3} mol dm^{-3}.
Using an estimated association constant of 1.03×10^4 mol^{-1} dm^3 and an ion-size parameter of å = 0.68 nm with which to calculate the mean ionic activity coefficient from the extended Debye-Hückel equation, they obtained for the standard Gibbs free energy of solution:
ΔG°_s = 8.78 kcal mol^{-1}
= 36.8 kJ mol^{-1} (compiler).
The solubility (ion-activity) product of Bu_4NBPh_4 can be calculated from the relationship:
$\Delta G^\circ_s = -RT \ln K^\circ_{s0}$, yielding pK°_{s0} = 6.437, where K°_{s0} units are mol^2 dm^{-6} (compiler).

AUXILIARY INFORMATION

METHOD/APPARATUS/PROCEDURE:

Evaporation and weighing. Saturated solutions prepared by shaking the suspensions for several days at 25°C. No solvate was detected. Method of temperature control was not specified.

SOURCE AND PURITY OF MATERIALS:

The solvent was shaken with anhydrous K_2CO_3, passed through a column of basic activated alumina into distillation flask and fractionated under N_2 through a 3-foot column. At least 10% of distillate was rejected, the rest collected over freshly activated molecular sieve. Bu_4NBPh_4 was recrystallized from aqueous acetone and vacuum dried at 60-80°C for several days.

ESTIMATED ERROR:

Precision of 0.1 kcal mol^{-1} in ΔG°_s.

REFERENCES:

COMPONENTS:

(1) Tetra-n-butylammonium tetraphenylborate (1-); $C_{40}H_{56}BN$; [15522-59-5]

(2) 1,2-Dichloroethane; $C_2H_4Cl_2$; [107-06-2]

ORIGINAL MEASUREMENTS:

Abraham, M. H. Danil de Namor, A. F. *J. Chem. Soc. Faraday Trans. 1* 1976, *72*, 955-62.

VARIABLES:

One temperature: 25°C

PREPARED BY:

Orest Popovych

EXPERIMENTAL VALUES:

The authors reported the solubility of Bu_4NBPh_4 in 1,2-dichloroethane as:
2.24×10^{-1} mol dm^{-3}.
Using an association constant of 1.715×10^3 mol^{-1} dm^3 (1) and an ion-size parameter of $\mathring{a} = 0.68$ nm with which to calculate the mean ionic activity coefficient from the extended Debye-Hückel equation, they obtained for the standard Gibbs free energy of solution:
$\Delta G_s^\circ = 5.76$ kcal mol^{-1}
$= 24.1$ kJ mol^{-1} (compiler).
The solubility (ion-activity) product of Bu_4NBPh_4 can be calculated from the relationship:
$\Delta G_s^\circ = -RT \ln K_{s0}^\circ$, yielding $pK_{s0}^\circ = 4.223$, where K_{s0}° units are mol^2 dm^{-6} (compiler).

AUXILIARY INFORMATION

METHOD/APPARATUS/PROCEDURE:

Evaporation and weighing. Saturated solutions prepared by shaking the suspensions for several days at 25°C. No solvate was detected. Method of temperature control was not specified.

SOURCE AND PURITY OF MATERIALS:

The solvent was shaken with anhydrous K_2CO_3, passed through a column of basic activated alumina into distillation flask and fractionated under N_2 through a 3-foot column. At least 10% of distillate was rejected, the rest collected over freshly activated molecular sieve. Bu_4NBPh_4 was recrystallized from aqueous acetone and vacuum dried at 60-80°C for several days.

ESTIMATED ERROR:

Precision of 0.1 kcal mol^{-1} in ΔG_s°.

REFERENCES:

(1) D'Aprano, A.; Fuoss, R. M. *J. Solution Chem.* 1975, *4*, 175.

COMPONENTS:

(1) Tetra-n-butylammonium tetraphenylborate (1-); $C_{40}H_{56}BN$; [15522-59-5]

(2) Methanol; CH_4O; [67-56-1]

ORIGINAL MEASUREMENTS:

Popovych, O.; Friedman, R. M. *J. Phys. Chem.* 1966, *70*, 1671-3.

VARIABLES:

One temperature: 25.00°C

PREPARED BY:

Orest Popovych

EXPERIMENTAL VALUES:

The authors report the solubility of tetra-n-butylammonium tetraphenylborate ($Bu_4N\ BPh_4$) in methanol as

$$C = 2.58 \times 10^{-3}\ \text{mol dm}^{-3}.$$

The solubility product, K°_{s0}, was calculated by the authors as $(C\alpha y_\pm)^2$ where the degree of dissociation α was obtained using a literature value of the ion-pair dissociation constant K_A (1) and the following expression:

$$\alpha = \frac{-1 + (1 + 4K_ACy_\pm^2)^{1/2}}{2K_ACy_\pm^2}$$

. The mean ionic activity coefficient $y_\pm$ was estimated from the Debye-Hückel equation in the form:

$$-\log y_\pm^2 = \frac{3.803\ (C\alpha)^{1/2}}{1 + 0.5099\ \mathring{a}(C\alpha)^{1/2}}$$ using $\mathring{a} = 0.7$ nm.

The above calculations yielded $\alpha = 0.940$ and $y_\pm^2 = 0.786$ from which $K^\circ_{s0} = 4.62 \times 10^{-6}\ \text{mol}^2\ \text{dm}^{-6}$ (compiler's units).
The molar absorption coefficients of the tetraphenylborate ion used to calculate the solubilities were reported to be 3.00×10^3 and 2.12×10^3 $dm^3(cm\ mol)^{-1}$ at 266 and 274 nm, respectively.

AUXILIARY INFORMATION

METHOD/APPARATUS/PROCEDURE:

UV spectrophotometry using a Cary Model 14 spectrophotometer. Saturartion achieved by shaking the salt suspensions for 2 weeks in water-jacketed flasks. Solutions filtered and analyzed at 266 and 274 nm. All solutions and containers were deaerated.

SOURCE AND PURITY OF MATERIALS:

Source and purification of methanol the same as described in the compilation for potassium tetraphenylborate in methanol. $Bu_4N\ BPh_4$ was synthesized and purified as described in the literature (1).

ESTIMATED ERROR:

Not specified, but the precision is known to be about ±1% for solubility.
Temperature: ±0.01°C.

REFERENCES:

(1) Coplan, M. A.; Fuoss, R. M. *J. Phys. Chem.* 1964, *68*, 1177.
(2) Accascina, F.; Petrucci, S.; Fuoss, R. M. *J. Am. Chem. Soc.* 1959, *81*, 1301.

COMPONENTS:

(1) Tetra-n-butylammonium tetraphenylborate (1-); $C_{40}H_{56}BN$; [15522-59-5]

(2) 1-Methyl-2-pyrrolidinone (N-methyl-2-pyrrolidone) C_5H_9NO; [872-50-4]

ORIGINAL MEASUREMENTS:

Virtanen, P. O. I.; Kerkelä, R. *Suom. Kemistil.* 1969, *B49*, 29-33.

VARIABLES:

Two temperatures: 25.00°C and 45.00°C.

PREPARED BY:

Orest Popovych

EXPERIMENTAL VALUES:

The solubility of n-Bu_4NBPh_4 in N-methyl-2-pyrrolidone was reported to be 0.964 mol dm^{-3} at 25°C and 1.08 mol dm^{-3} at 45°C.

The corresponding solubility product at 25°C, calculated as the square of the solubility, was reported in the form $pK_{s0} = 0.03$, where K_{s0} units are $mol^2\ dm^{-6}$. The solubility product at 45°C was not reported.

AUXILIARY INFORMATION

METHOD/APPARATUS/PROCEDURE:

The suspensions were shaken in thermostatted water-jacketed flasks for one day at 50°C, followed by one day at 25°C or 45°C, respectively. Saturated solutions were analyzed gravimetrically after precipitation of $KBPh_4$ or NH_4BPh_4 from aliquots in aqueous solution.

SOURCE AND PURITY OF MATERIALS:

N-methyl-2-pyrrolidone (General Aniline & Film Co.) was purified as in the literature (1). n-Bu_4NBPh_4 was prepared by metathesis of aqueous $NaBPh_4$ with methanolic n-Bu_4NI, followed by double recrystallization from an acetone-water mixture and drying in vacuo.

ESTIMATED ERROR:

Not specified.
Temperature control: ±0.02°C

REFERENCES:

(1) Virtanen, P. O. I. *Suom. Kemistil.* 1966, *B39*, 257.

COMPONENTS:	ORIGINAL MEASUREMENTS:
(1) Tetraethylammonium tetraphenylborate (1-); $C_{32}H_{40}BN$; [12099-10-4] (2) 1,1-Dichloroethane; $C_2H_4Cl_2$; [75-34-3]	Abraham, M. H.; Danil de Namor, A. F. *J. Chem. Soc. Faraday Trans. 1* 1976, *72*, 955-62.
VARIABLES:	PREPARED BY:
One temperature: 25°C	Orest Popovych

EXPERIMENTAL VALUES:

The authors reported the solubility of Et_4NBPh_4 in 1,1-dichloroethane as: 4.14×10^{-4} mol dm^{-3}.
Using an estimated association constant of 1.50×10^4 mol^{-1} dm^3 and an ion-size parameter $\mathring{a} = 0.64$ nm with which to calculate the mean ionic activity coefficient from the extended Debye-Hückel equation, they obtained for the standard Gibbs free energy of solution:

$$\Delta G_s^\circ = 10.64 \text{ kcal mol}^{-1} = 44.54 \text{ kJ mol}^{-1} \text{ (compiler)}.$$

The solubility (ion-activity) product of Et_4NBPh_4 can be calculated from the relationship: $\Delta G_s^\circ = -RT \ln K_{s0}^\circ$, yielding $pK_{s0}^\circ = 7.801$, where K_{s0}° units are mol^2 dm^{-6} (compiler).

AUXILIARY INFORMATION

METHOD/APPARATUS/PROCEDURE:	SOURCE AND PURITY OF MATERIALS:
Evaporation and weighing. Saturated solutions prepared by shaking the suspensions for several days at 25°C. No solvate was detected. Method of temperature control was not specified.	The solvent was shaken with anhydrous K_2CO_3, passed through a column of basic activated alumina into distillation flask and fractionated under N_2 through a 3-foot column. At least 10% of distillate was rejected, the rest collected over freshly activated molecular sieve. Et_4NBPh_4 was recrystallized from aqueous acetone and vacuum dried for several days at 60-80°C.
	ESTIMATED ERROR:
	Precision of 0.1 kcal mol^{-1} in ΔG_s°.
	REFERENCES:

COMPONENTS:	ORIGINAL MEASUREMENTS:
(1) Tetraethylammonium tetraphenylborate (1-); $C_{32}H_{40}BN$; [12099-10-4] (2) 1,2-Dichloroethane; $C_2H_4Cl_2$; [107-06-2]	Abraham, M. H.; Danil de Namor, A. F. *J. Chem. Soc. Faraday Trans. 1* 1976, *72*, 955-62.
VARIABLES:	PREPARED BY:
One temperature: 25°C	Orest Popovych

EXPERIMENTAL VALUES:

The authors reported the solubility of Et_4NBPh_4 in 1,2-dichloroethane as: 8.40×10^{-3} mol dm^{-3}.
Using an estimated association constant of 2.50×10^3 mol^{-1} dm^3 and an ion-size parameter of $\mathring{a}$ = 0.64 nm with which to calculate the mean ionic activity coefficient from the extended Debye-Hückel equation, they obtained for the standard Gibbs free energy of solution:

$$\Delta G_s^\circ = 7.88 \text{ kcal mol}^{-1} = 33.0 \text{ kJ mol}^{-1} \text{ (compiler)}.$$

The solubility (ion-activity) product of Et_4NBPh_4 can be calculated from the relationship: $\Delta G_s^\circ = -RT \ln K_{s0}^\circ$, yielding $pK_{s0}^\circ = 5.777$, where K_{s0}° units are mol^2 dm^{-6} (compiler).

AUXILIARY INFORMATION

METHOD/APPARATUS/PROCEDURE:	SOURCE AND PURITY OF MATERIALS:
Evaporation and weighing. Saturated solutions prepared by shaking the suspensions for several days at 25°C. No solvate was detected. Method of temperature control was not specified.	The solvent was shaken with anhydrous K_2CO_3, passed through a column of basic activated alumina into distillation flask and fractionated under N_2 through a 3-foot column. At least 10% of distillate was rejected, the rest collected over freshly activated molecular sieve. Et_4NBPh_4 was recrystallized from aqueous acetone and vacuum dried for several days at 60-80°C.
	ESTIMATED ERROR:
	Precision of 0.1 kcal mol^{-1} in ΔG_s°.
	REFERENCES:

COMPONENTS:

(1) Tetraethylammonium tetraphenylborate (1-); $C_{32}H_{40}BN$; [12099-10-4]

(2) 1-Propanol; C_3H_8O; [71-23-8]

ORIGINAL MEASUREMENTS:

Abraham, M. H.; Danil de Namor, A. F. *J. Chem. Soc. Faraday Trans. 1* 1978, *74*, 2101-10.

VARIABLES:

One temperature: 25°C

PREPARED BY:

Orest Popovych

EXPERIMENTAL VALUES:

The solubility of Et_4NBPh_4 in 1-propanol was reported as:

4.03×10^{-4} mol dm^{-3}.

Using an estimated association constant of 800 mol^{-1} dm^3 (1) and an ion-size parameter of 0.65 nm with which to calculate the mean ionic activity coefficient from the extended Debye-Hückel equation, the authors obtained for the standard Gibbs free energy of solution:

ΔG_s° = 9.66 kcal mol^{-1}, which is 40.42 kJ mol^{-1} (compiler).

The solubility (ion-activity) product of Et_4NBPh_4 can be calculated from the relationship $\Delta G_s^\circ = -RT \ln K_{s0}^\circ$, yielding $pK_{s0}^\circ = 7.082$, where K_{s0}° units are mol^2 dm^{-6} (compiler).

AUXILIARY INFORMATION

METHOD/APPARATUS/PROCEDURE:

Evaporation and weighing. Saturated solutions were prepared by shaking the suspensions for several days. The solvent contained no involatile material and the solute formed no solvate. Method of temperature control was not specified.

SOURCE AND PURITY OF MATERIALS:

The purification of the solvent was described in the literature (2). Et_4NBPh_4 was recrystallized from aqueous acetone and vacuum dried for several days at 60-80°C.

ESTIMATED ERROR:

Precision of 0.15 kcal mol^{-1} in ΔG_s°.

REFERENCES:

(1) Abraham, M. H.; Lee, W. H.; Wheaton, R. S. *J. Solution Chem.* in press.

(2) Abraham, M. H.; Danil de Namor, A. F.; Schulz, R. A. *J. Solution Chem.* 1977, *6*, 491.

COMPONENTS:	EVALUATOR:
(1) Tetramethylammonium tetraphenylborate (1-); $C_{28}H_{32}BN$; [15525-13-0] (2) Water; H_2O; [7732-18-5]	Orest Popovych, Department of Chemistry, City University of New York, Brooklyn College, Brooklyn, N. Y. 11210, U. S. A. March 1980

CRITICAL EVALUATION:

Although the solubility of tetramethylammonium tetraphenylborate in water was reported from two different studies (1,2), there is no doubt as to which of them is the more reliable. Thus, the solubility reported by Zeidler (1) as 0.05% is not only ambiguous and expressed to one significant digit, but also was determined by an unspecified analytical method at an unspecified (presumably room) temperature. On the other hand, the value reported by Howick and Pflaum (2) as 4.3×10^{-5} mol dm^{-3} was obtained by uv-spectrophotometry at 298 K under conditions where saturation was ascertained.

There are unfortunately two drawbacks in the latter study as well. First, the extent to which the temperature was controlled is not specified. Second, it is not clear which values of the absorption coefficients ε_{max} for the tetraphenylborate ion were used to calculate the solubilities. This question arises because of the fact that in an earlier study Pflaum and Howick (3) used ε_{max} values characteristic of acetonitrile solutions to calculate solubilities in aqueous solutions from their absorption data. The molar ε_{max} values used by Pflaum and Howick (3) were 3.225×10^3 and 2.110×10^3 at 266 nm and 274 nm, respectively, while the corresponding values reported for aqueous solutions are 3.25×10^3 and 2.06×10^3, respectively (4). (All molar absorption coefficients are in the units of dm^3 (cm mol)$^{-1}$). Thus, if Howick and Pflaum (2) averaged the solubility values determined at the two wavelengths, they benefitted from a compensation of errors, which in the case of the solubility of $KBPh_4$ in water (3) led to excellent agreement with other literature data (see critical evaluation for $KBPh_4$ in water). However, one cannot be certain that a similar compensation of errors was involved in the case of tetramethylammonium tetraphenylborate. Consequently, the solubility of 4.3×10^{-5} mol dm^{-3} at 298 K must be regarded as a highly tentative value.

REFERENCES:

1. Zeidler, L. *Hoppe-Seyler's Z. Physiol. Chem.* 1952, *291*, 177.
2. Howick, L. C.; Pflaum, R. T. *Anal. Chim. Acta* 1958, *19*, 343.
3. Pflaum, R. T.; Howick, L. C. *Anal Chem.* 1956, *28*, 1542.
4. Popovych, O.; Friedman, R. M. *J. Phys. Chem.* 1966, *70*, 1671.

COMPONENTS:

(1) Tetramethylammonium tetraphenylborate (1-); $C_{28}H_{32}BN$; [15525-13-0]

(2) Water; H_2O; [7732-18-5]

ORIGINAL MEASUREMENTS:

Zeidler, L. *Hoppe-Seyler's Z. Physiol. Chem.* 1952, *291*, 177-8.

VARIABLES:

Presumably room temperature

PREPARED BY:

Orest Popovych

EXPERIMENTAL VALUES:

The solubility of tetramethylammonium tetraphenylborate was reported as 0.05%, probably meaning 0.05 g in 100 cm^3 of saturated solution. If this interpretation is correct, the solubility corresponds to $1._3 \times 10^{-3}$ mol dm^{-3} (compiler).

AUXILIARY INFORMATION

METHOD/APPARATUS/PROCEDURE:

Nothing specified.

SOURCE AND PURITY OF MATERIALS:

The salt was prepared by reacting tetramethylammonium hydroxide (it isn't clear whether its solution was adjusted to neutrality or slight acidity as in the case of other bases in this study) with a freshly prepared solution of $NaBPh_4$ ("Kalignost" from Heyl & Co.). Analysis of the product yielded 3.28% N, as compared to 3.56% theoretical.

ESTIMATED ERROR:

Nothing specified.

REFERENCES:

COMPONENTS:

(1) Tetramethylammonium tetraphenylborate (1-); $C_{28}H_{32}BN$; [15525-13-0]

(2) Water, H_2O; [7732-18-5]

ORIGINAL MEASUREMENTS:

Howick, L. C.; Pflaum, R. T.
Anal. Chim. Acta 1958, *19*, 343-7.

VARIABLES:

One temperature: 25°C

PREPARED BY:

Orest Popovych

EXPERIMENTAL VALUES:

The solubility of tetramethylammonium tetraphenylborate in water was reported as:

$$4.3 \times 10^{-5} \text{ mol dm}^{-3}.$$

AUXILIARY INFORMATION

METHOD/APPARATUS/PROCEDURE:

Saturated solutions were prepared both by agitating the suspensions at 25°C continuously and by agitating them first for a 0.5 hr at 40-50°C and then cooling to 25°C. When equilibrium was attained, the filtered solutions were analyzed for the BPh_4^- anion by UV spectrophotometry, using a Cary Model 11 recording spectrophotometer. The method of temperature control was not stated.

SOURCE AND PURITY OF MATERIALS:

See the compilation for NH_4BPh_4 in water based on the same reference. The amine hydrochloride used to prepare the tetraphenylborate was an Eastman White Label product.

ESTIMATED ERROR:

Nothing specified.

REFERENCES:

COMPONENTS:

(1) Tetramethylammonium tetraphenylborate (1-); $C_{28}H_{32}BN$; [15525-13-0]

(2) 1,1-Dichloroethane; $C_2H_4Cl_2$; [75-34-3]

ORIGINAL MEASUREMENTS:

Abraham, M. H.; Danil de Namor, A. F. *J. Chem. Soc. Faraday Trans. 1* 1976, *72*, 955-62.

VARIABLES:

One temperature: 25°C

PREPARED BY:

Orest Popovych

EXPERIMENTAL VALUES:

The authors reported the solubility of Me_4NBPh_4 in 1,1-dichloroethane as:
7.21×10^{-5} mol dm^{-3}.
Using an estimated association constant of 7.44×10^4 mol^{-1} dm^3 and an ion-size parameter of å = 0.62 nm with which to calculate the mean ionic activity coefficient from the extended Debye-Hückel equation, they obtained for the standard Gibbs free energy of solution:

$$\Delta G_s^\circ = 12.59 \text{ kcal mol}^{-1} = 52.70 \text{ kJ mol}^{-1} \text{ (compiler).}$$

The solubility (ion-activity) product of Me_4NBPh_4 can be calculated from the relationship:

$\Delta G_s^\circ = -RT \ln K_{s0}^\circ$, yielding $pK_{s0}^\circ = 9.230$, where K_{s0}° units are mol^2 dm^{-6} (compiler).

AUXILIARY INFORMATION

METHOD/APPARATUS/PROCEDURE:

Evaporation and weighing. Saturated solutions prepared by shaking the suspensions for several days at 25°C. No solvate was detected. Method of temperature control was not specified.

SOURCE AND PURITY OF MATERIALS:

The solvent was shaken with anhydrous K_2CO_3, passed through a column of basic activated alumina into distillation flask and fractionated under N_2 through a 3-foot column. At least 10% of distillate was rejected, the rest collected over freshly activated molecular sieve. Me_4NBPh_4 was recrystallized from acetone and vacuum dried at 60-80°C for several days.

ESTIMATED ERROR:

Precision of 0.1 kcal mol^{-1} in ΔG_s°.

REFERENCES:

COMPONENTS:

(1) Tetramethylammonium tetraphenylborate (1-); $C_{28}H_{32}BN$; [15525-13-0]

(2) 1,2-Dichloroethane; $C_2H_4Cl_2$; [107-06-2]

ORIGINAL MEASUREMENTS:

Abraham, M. H.; Danil de Namor A. F. *J. Chem. Soc. Faraday Trans. 1* 1976, *72*, 955-62.

VARIABLES:

One temperature: 25°C

PREPARED BY:

Orest Popovych

EXPERIMENTAL VALUES:

The authors reported the solubility of Me_4NBPh_4 in 1,2-dichloroethane as: 4.12×10^{-4} mol dm^{-3}.
Using an estimated association constant of 1.24×10^4 mol^{-1} dm^3 and an ion-size parameter of å = 0.62 nm with which to calculate the mean ionic activity coefficients from the extended Debye-Hückel equation, they obtained for the standard Gibbs free energy of solution:

$$\Delta G_s^\circ = 10.56 \text{ kcal mol}^{-1} = 44.20 \text{ kJ mol}^{-1} \text{ (compiler)}.$$

The solubility (ion-activity) product of Me_4NBPh_4 can be calculated from the relationship: $\Delta G_s^\circ = -RT \ln K_{s0}^\circ$, yielding $pK_{s0}^\circ = 7.742$, where K_{s0}° units are mol^2 dm^{-6} (compiler).

AUXILIARY INFORMATION

METHOD/APPARATUS/PROCEDURE:

Evaporation and weighing. Saturated solutions prepared by shaking the suspensions for several days at 25°C. No solvate was detected. Method of temperature control was not specified.

SOURCE AND PURITY OF MATERIALS:

The solvent was shaken with anhydrous K_2CO_3, passed through a column of basic activated alumina into distillation flask and fractionated under N_2 through a 3-foot column. At least 10% of distillate was rejected, the rest collected over freshly activated molecular sieve. Me_4NBPh_4 was recrystallized from acetone and vacuum dried for several days at 60-80°C.

ESTIMATED ERROR:

Precision of 0.1 kcal mol^{-1} in ΔG_s°.

REFERENCES:

COMPONENTS:

(1) Tetramethylammonium tetraphenylborate (1-); $C_{28}H_{32}BN$; [15525-13-0]

(2) 1-Propanol; C_3H_8O; [71-23-8]

ORIGINAL MEASUREMENTS:

Abraham, M. H.; Danil de Namor, A. F. *J. Chem. Soc. Faraday Trans. 1* 1978, *74*, 2101-10.

VARIABLES:

One temperature: 25°C

PREPARED BY:

Orest Popovych

EXPERIMENTAL VALUES:

The solubility of Me_4NBPh_4 in 1-propanol was reported as:

1.07×10^{-4} mol dm^{-3}.

Using an estimated association constant of 1300 mol^{-1} dm^3 (1) and an ion-size parameter of 0.60 nm with which to calculate the mean ionic activity coefficient from the extended Debye-Hückel equation, the authors obtained for the standard Gibbs free energy of solution:

ΔG°_s = 11.06 kcal mol^{-1}, which is 46.28 kJ mol^{-1} (compiler).

The solubility (ion-activity) product of Me_4NBPh_4 can be calculated from the relationship $\Delta G^\circ_s = -RT \ln K^\circ_{s0}$, yielding pK°_{s0} = 8.109, where K°_{s0} units are mol^2 dm^{-6} (compiler).

AUXILIARY INFORMATION

METHOD/APPARATUS/PROCEDURE:

Evaporation and weighing. Saturated solutions were prepared by shaking the suspensions for several days. The solvent contained no involatile material and the solute formed no solvate. Method of temperature control was not specified.

SOURCE AND PURITY OF MATERIALS:

The purification of the solvent was described in the literature (2). Me_4NBPh_4 was recrystallized from acetone and vacuum dried for several days at 60-80°C.

ESTIMATED ERROR:

Precision of 0.15 kcal mol^{-1} in ΔG°_s.

REFERENCES:

(1) Abraham, M. H.; Lee, W. H.; Wheaton, R. S. *J. Solution Chem.*, in press.

(2) Abraham, M. H.; Danil de Namor, A. F.; Schulz, R. A. *J. Solution Chem.* 1977, *6*, 491.

COMPONENTS:

(1) Tetrapropylammonium tetraphenylborate (1-); $C_{36}H_{48}BN$; [15556-39-5]

(2) 1,1-Dichloroethane; $C_2H_4Cl_2$; [75-34-3]

ORIGINAL MEASUREMENTS:

Abraham, M. H.; Danil de Namor, A. F. *J. Chem. Soc. Faraday Trans. 1* 1976, *72*, 955-62.

VARIABLES:

One temperature: 25°C

PREPARED BY:

Orest Popovych

EXPERIMENTAL VALUES:

The authors reported the solubility of Pr_4NBPh_4 in 1,1-dichloroethane as: 2.61×10^{-3} mol dm^{-3}.
Using an estimated association constant of 1.26×10^4 mol^{-1} dm^3 and an ion-size parameter $\mathring{a}$ = 0.66 nm with which to calculate the mean ionic activity coefficient from the extended Debye-Hückel equation, they obtained for the standard Gibbs free energy of solution:

$$\Delta G_s^\circ = 9.30 \text{ kcal mol}^{-1} = 38.9 \text{ kJ mol}^{-1} \text{ (compiler)}.$$

The solubility (ion-activity) product of Pr_4NBPh_4 can be calculated from the relationship: $\Delta G_s^\circ = -RT \ln K_{s0}^\circ$, yielding $pK_{s0}^\circ = 6.818$, where K_{s0}° units are mol^2 dm^{-6} (compiler).

AUXILIARY INFORMATION

METHOD/APPARATUS/PROCEDURE:

Evaporation and weighing. Saturated solutions were prepared by shaking the suspensions for several days at 25°C. No solvate was detected. Method of temperature control was not specified.

SOURCE AND PURITY OF MATERIALS:

The solvent was shaken with anhydrous K_2CO_3, passed through a column of basic activated alumina into distillation flask and fractionated under N_2 through a 3-foot column. At least 10% of distillate was rejected, the rest collected over freshly activated molecular sieve. Pr_4NBPh_4 was recrystallized from aqueous acetone and vacuum dried for several days at 60-80°C.

ESTIMATED ERROR:

Precision of 0.1 kcal mol^{-1} in ΔG_s°.

REFERENCES:

COMPONENTS:

(1) Tetrapropylammonium tetraphenylborate (1-); $C_{36}H_{48}BN$; [15556-39-5]

(2) 1,2-Dichloroethane; $C_2H_4Cl_2$; [107-06-2]

ORIGINAL MEASUREMENTS:

Abraham, M. H.; Danil de Namor A. F. *J. Chem. Soc. Faraday Trans. 1* 1976, *72*, 955-62.

VARIABLES:

One temperature: 25°C

PREPARED BY:

Orest Popovych

EXPERIMENTAL VALUES:

The authors reported the solubility of Pr_4NBPh_4 in 1,2-dichloroethane as: 1.01×10^{-1} mol dm^{-3}.
Using an estimated association constant of 2.10×10^3 mol^{-1} dm^3 and an ion-size parameter of $\mathring{a}$ = 0.66 nm with which to calculate the mean ionic activity coefficient from the extended Debye-Hückel equation, they obtained for the standard Gibbs free energy solution:

$$G_s^\circ = 6.33 \text{ kcal mol}^{-1} = 26.5 \text{ kJ mol}^{-1} \text{ (compiler)}.$$

The solubility (ion-activity) product of Pr_4NBPh_4 can be calculated from relationship: $\Delta G_s^\circ = -RT \ln K_{s0}^\circ$, yielding $pK_{s0}^\circ = 4.640$, where K_{s0}° units are mol^2 dm^{-6} (compiler).

AUXILIARY INFORMATION

METHOD/APPARATUS/PROCEDURE:

Evaporation and weighing. Saturated solutions prepared by shaking the suspensions for several days at 25°C. No solvate was detected. Method of temperature control was not specified.

SOURCE AND PURITY OF MATERIALS:

The solvent was shaken with anhydrous K_2CO_3, passed through a column of basic activated alumina into distillation flask and fractionated under N_2 through a 3-foot column. At least 10% of distillate was rejected, the rest collected over freshly activated molecular sieve. Pr_4NBPh_4 was recrystallized from aqueous acetone and vacuum dried for several days at 60-80°C.

ESTIMATED ERROR:

Precision of 0.1 kcal mol^{-1} in ΔG_s°.

REFERENCES:

COMPONENTS:

(1) Tetrapropylammonium tetraphenylborate (1-); $C_{36}H_{48}BN$; [15556-39-5]

(2) 1-Propanol; C_3H_8O; [71-23-8]

ORIGINAL MEASUREMENTS:

Abraham, M. H.; Danil de Namor, A. F. *J. Chem. Soc. Faraday Trans. 1* 1978, *74*, 2101-10.

VARIABLES:

One temperature: 25°C

PREPARED BY:

Orest Popovych

EXPERIMENTAL VALUES:

The solubility of Pr_4NBPh_4 in 1-propanol was reported as:

6.40×10^{-4} mol dm^{-3}.

Using an estimated association constant of 670 mol^{-1}dm^{3} (1) and an ion-size parameter of 0.50 nm with which to calculate the mean ionic activity coefficient from the extended Debye-Hückel equation, the authors obtained for the standard Gibbs free energy of solution:

ΔG°_s = 9.19 kcal mol^{-1}, which is 38.45 kJ mol^{-1} (compiler).

The solubility (ion-activity) product of Pr_4NBPh_4 can be calculated from the relationship $\Delta G^\circ_s = -RT \ln K^\circ_{s0}$, yielding $pK^\circ_{s0} = 6.738$, where K°_{s0} units are mol^2 dm^{-6} (compiler).

AUXILIARY INFORMATION

METHOD/APPARATUS/PROCEDURE:

Evaporation and weighing. Saturated solutions were prepared by shaking the suspensions for several days. The solvent contained no involatile material and the solute formed no solvate. Method of temperature control was not specified.

SOURCE AND PURITY OF MATERIALS:

The purification of the solvent was described in the literature (2). Pr_4NBPh_4 was recrystallized from aqueous acetone and vacuum dried for several days at 60-80°C.

ESTIMATED ERROR:

Precision of 0.15 kcal mol^{-1} in ΔG°_s.

REFERENCES:

(1) Abraham, M. H.; Lee, W. H.; Wheaton, R. S. *J. Solution Chem.* in press.

(2) Abraham, M. H.; Danil de Namor, A. F.; Schulz, R. A. *J. Solution Chem.* 1977, *6*, 491.

COMPONENTS:

(1) Trimethylammonium tetraphenylborate (1-); $C_{27}H_{30}BN$; [51016-92-3]

(2) Water; H_2O; [7732-18-5]

ORIGINAL MEASUREMENTS:

Howick, L. C.; Pflaum, R. T.
Anal. Chim. Acta 1958, *19*, 343-7.

VARIABLES:

One temperature: 25°C

PREPARED BY:

Orest Popovych

EXPERIMENTAL VALUES:

The solubility of trimethylammonium tetraphenylborate in water was reported as:

$$3.87 \times 10^{-4} \text{ mol dm}^{-3}.$$

For a critical evaluation of the data from this study, see the evaluation for NH_4BPh_4 in water.

AUXILIARY INFORMATION

METHOD/APPARATUS/PROCEDURE:

Saturated solutions were prepared both by agitating the suspensions continuously at 25°C and by agitating them first for a 0.5 hr at 40-50°C and then cooling to 25°C. When equilibrium was attained, the filtered solutions were analyzed for the BPh_4^- anion by UV spectrophotometry, using a Cary Model 11 recording spectrophotometer. The method of temperature control was not stated.

SOURCE AND PURITY OF MATERIALS:

See the compilation for NH_4BPh_4 in water based on the same reference. The amine hydrochloride used to prepare the tetraphenylborate was an Eastman White Label product.

ESTIMATED ERROR:

Nothing specified.

REFERENCES:

COMPONENTS:

(1) Tris(o-phenanthroline)ruthenium (II) tetraphenylborate (1-); $RuC_{84}H_{64}B_2N_6$;

(2) Organic Solvents

EVALUATOR:

Orest Popovych, Department of Chemistry, City University of New York, Brooklyn College, Brooklyn, N. Y. 11210, U. S. A.
October 1979

CRITICAL EVALUATION:

There is only one published value for the solubility of tris(o-phenanthroline)ruthenium(II) tetraphenylborate in each of the 22 organic solvents at 298 K as given in the compilations that follow (1). Unfortunately, the accuracy of these solubility values is probably adversely affected by two shortcomings: 1) The molar absorption coefficient of the cation, which was used to calculate the solubiluty in different organic solvents was characteristic of aqueous solutions and 2) the precision of the temperature control was not specified. Consequently, the reported solubility values must be disignated as no better than tentative.

REFERENCE:

1. Takamatsu, T. *Bull. Chem. Soc. Japan* 1974, *47*, 1285.

COMPONENTS:

(1) Tris(o-phenanthroline)ruthenium (II) tetraphenylborate (1-); $RuC_{84}H_{64}B_2N_6$;

(2) Benzyl alcohol; C_7H_8O; [100-51-6]

ORIGINAL MEASUREMENTS:

Takamatsu, T. *Bull. Chem. Soc. Japan* 1974, *47*, 1285-6.

VARIABLES:

One temperature: 25°C

PREPARED BY:

Orest Popovych

EXPERIMENTAL VALUES:

The solubility of $Ru(phen)_3(BPh_4)_2$ in benzyl alcohol was reported as:

2.68×10^{-4} mol dm^{-3}.

AUXILIARY INFORMATION

METHOD/APPARATUS/PROCEDURE:

Spectrophotometry. For details see compilation for $Ru(phen)_3(BPh_4)_2$ in acetone based on the same lit. reference.

SOURCE AND PURITY OF MATERIALS:

See compilation for $Ru(phen)_3(BPh_4)_2$ in acetone based on the same lit. reference.

ESTIMATED ERROR:

Nothing specified.

REFERENCES:

COMPONENTS:

(1) Tris(o-phenanthroline)ruthenium (II) tetraphenylborate (1-); $RuC_{84}H_{64}B_2N_6$;

(2) 2-Butanone (ethyl methyl ketone); C_4H_8O; [78-93-3]

ORIGINAL MEASUREMENTS:

Takamatsu, T. *Bull. Chem. Soc. Japan* 1974, *47*, 1285-6.

VARIABLES:

One temperature: 25°C

PREPARED BY:

Orest Popovych

EXPERIMENTAL VALUES:

The solubility of $Ru(phen)_3(BPh_4)_2$ in 2-butanone was reported as:

2.45×10^{-3} mol dm^{-3}.

AUXILIARY INFORMATION

METHOD/APPARATUS/PROCEDURE:

Spectrophotometry. For details see compilation for $Ru(phen)_3(BPh_4)_2$ in acetone based on the same lit. reference.

SOURCE AND PURITY OF MATERIALS:

See compilation for $Ru(phen)_3(BPh_4)_2$ in acetone based on the same lit. reference.

ESTIMATED ERROR:

Nothing specified.

REFERENCES:

COMPONENTS:

(1) Tris(o-phenanthroline)ruthenium (II) tetraphenylborate (1-); $RuC_{84}H_{64}B_2N_6$;

(2) sec-Butyl alcohol (2-butanol); $C_4H_{10}O$; [78-92-2]

ORIGINAL MEASUREMENTS:

Takamatsu, T. *Bull. Chem. Soc. Japan* 1974, *47*, 1285-6.

VARIABLES:

One temperature: 25°C

PREPARED BY:

Orest Popovych

EXPERIMENTAL VALUES:

The solubility of $Ru(phen)_3(BPh_4)_2$ in 2-butanol was reported as:

2.05×10^{-6} mol dm^{-3}.

AUXILIARY INFORMATION

METHOD/APPARATUS/PROCEDURE:

Spectrophotometry. For details see compilation for $Ru(phen)_3(BPh_4)_2$ in acetone based on the same lit. reference.

SOURCE AND PURITY OF MATERIALS:

See compilation for $Ru(phen)_3(BPh_4)_2$ in acetone based on the same lit. reference.

ESTIMATED ERROR:

Nothing specified.

REFERENCES:

COMPONENTS:	ORIGINAL MEASUREMENTS:
(1) Tris(o-phenanthroline)ruthernium (II) tetraphenylborate (1-); $RuC_{84}H_{64}B_2N_6$; (2) n-Butyl ethanoate (n-butyl acetate); $C_6H_{12}O_2$; [123-86-4]	Takamatsu, T. *Bull. Chem. Soc. Japan* 1974, *47*, 1285-6.
VARIABLES: One temperature: 25°C	PREPARED BY: Orest Popovych

EXPERIMENTAL VALUES:

The solubility of $Ru(phen)_3(BPh_4)_2$ in n-butyl acetate was reported as:

5.94×10^{-7} mol dm^{-3}.

AUXILIARY INFORMATION

METHOD/APPARATUS/PROCEDURE:	SOURCE AND PURITY OF MATERIALS:
Spectrophotometry. For details see compilation for $Ru(phen)_3(BPh_4)_2$ in acetone based on the same lit. reference.	See compilation for $Ru(phen)_3(BPh_4)_2$ in acetone based on the same lit. reference.
	ESTIMATED ERROR: Nothing specified.
	REFERENCES:

COMPONENTS:

(1) Tris(o-phenanthroline)ruthenium (II) tetraphenylborate (1-); $RuC_{84}H_{64}B_2N_6$;

(2) Chlorobenzene; C_6H_5Cl; [108-90-7]

ORIGINAL MEASUREMENTS:

Takamatsu, T. *Bull Chem. Soc. Japan* 1974, *47*, 1285-6.

VARIABLES:

One temperature: 25°C

PREPARED BY:

Orest Popovych

EXPERIMENTAL VALUES:

The solubility of $Ru(phen)_3(BPh_4)_2$ in chlorobenzene was reported as:

7.03×10^{-6} mol dm^{-3}.

AUXILIARY INFORMATION

METHOD/APPARATUS/PROCEDURE:

Spectrophotometry. For details see compilation for $Ru(phen)_3(BPh_4)_2$ in acetone based on the same lit. reference.

SOURCE AND PURITY OF MATERIALS:

See compilation for $Ru(phen)_3(BPh_4)_2$ in acetone based on the same lit. reference.

ESTIMATED ERROR:

Nothing specified.

REFERENCES:

COMPONENTS:

(1) Tris(o-phenanthroline)ruthenium (II) tetraphenylborate (1-); $RuC_{84}H_{64}B_2N_6$;

(2) bis-2-Chloroethyl ether; $C_4H_8Cl_2O$; [111-44-4]

ORIGINAL MEASUREMENTS:

Takamatsu, T. *Bull. Chem. Soc. Japan* 1974, *47*, 1285-6.

VARIABLES:

One temperature: 25°C

PREPARED BY:

Orest Popovych

EXPERIMENTAL VALUES:

The solubility of $Ru(phen)_3(BPh_4)_2$ in bis-2-chloroethyl ether was reported as:

1.26×10^{-2} mol dm^{-3}.

AUXILIARY INFORMATION

METHOD/APPARATUS/PROCEDURE:

Spectrophotometry. For details see compilation for $Ru(phen)_3(BPh_4)_2$ in acetone based on the same lit. reference.

SOURCE AND PURITY OF MATERIALS:

See compilation for $Ru(phen)_3(BPh_4)_2$ in acetone based on the same lit. reference.

ESTIMATED ERROR:

Nothing specified.

REFERENCES:

COMPONENTS:	ORIGINAL MEASUREMENTS:
(1) Tris(o-phenanthroline)ruthenium (II) tetraphenylborate (1-); $RuC_{84}H_{64}B_2N_6$; (2) Chloroform; $CHCl_3$; [67-66-3]	Takamatsu, T. *Bull. Chem. Soc. Japan* 1974, *47*, 1285-6.
VARIABLES:	PREPARED BY:
One temperature: 25°C	Orest Popovych

EXPERIMENTAL VALUES:

The solubility of $Ru(phen)_3(BPh_4)_2$ in chloroform was reported as:

1.29×10^{-4} mol dm^{-3}.

AUXILIARY INFORMATION

METHOD/APPARATUS/PROCEDURE:

Spectrophotometry. For details see compilation for $Ru(phen)_3(BPh_4)_2$ in acetone based on the same lit. reference.

SOURCE AND PURITY OF MATERIALS:

See compilation for $Ru(phen)_3(BPh_4)_2$ in acetone based on the same lit. reference.

ESTIMATED ERROR:

Nothing specified.

REFERENCES:

COMPONENTS:

(1) Tris(o-phenanthroline)ruthenium (II) tetraphenylborate (1-); $RuC_{84}H_{64}B_2N_6$;

(2) 1,2-Dichloroethane; $C_2H_4Cl_2$; [107-06-2]

ORIGINAL MEASUREMENTS:

Takamatsu, T. *Bull. Chem. Soc. Japan* 1974, *47*, 1285-6.

VARIABLES:

One temperature: 25°C

PREPARED BY:

Orest Popovych

EXPERIMENTAL VALUES:

The solubility of $Ru(phen)_3(BPh_4)_2$ in 1,2-dichloroethane was reported as:

$$4.18 \times 10^{-4} \text{ mol dm}^{-3}.$$

AUXILIARY INFORMATION

METHOD/APPARATUS/PROCEDURE:

Spectrophotometry. For details see compilation for $Ru(phen)_3(BPh_4)_2$ in acetone based on the same lit. reference.

SOURCE AND PURITY OF MATERIALS:

See compilation for $Ru(phen)_3(BPh_4)_2$ in acetone based on the same lit. reference.

ESTIMATED ERROR:

Nothing specified.

REFERENCES:

COMPONENTS:

(1) Tris(o-phenanthroline)ruthenium (II) tetraphenylborate (1-); $RuC_{84}H_{64}B_2N_6$;

(2) 3,3-Dimethyl-2-butanone (methyl isobutyl ketone); $C_6H_{12}O$; [75-97-8]

ORIGINAL MEASUREMENTS:

Takamatsu, T. *Bull. Chem. Soc. Japan* 1974, *47*, 1285-6.

VARIABLES:

One temperature: 25°C

PREPARED BY:

Orest Popovych

EXPERIMENTAL VALUES:

The solubility of $Ru(phen)_3(BPh_4)_2$ in methyl isobutyl ketone was reported as:

$$6.92 \times 10^{-4} \text{ mol dm}^{-3}.$$

AUXILIARY INFORMATION

METHOD/APPARATUS/PROCEDURE:

Spectrophotometry. For details see compilation for $Ru(phen)_3(BPh_4)_2$ in acetone based on the same lit. reference.

SOURCE AND PURITY OF MATERIALS:

See compilation for $Ru(phen)_3(BPh_4)_2$ in acetone based on the same lit. reference.

ESTIMATED ERROR:

Nothing specified.

REFERENCES:

COMPONENTS:	ORIGINAL MEASUREMENTS:
(1) Tris(o-phenanthroline)ruthenium (II) tetraphenylborate (1-); $RuC_{84}H_{64}B_2N_6$; (2) Ethanol; C_2H_6O; [64-17-5]	Takamatsu, T. *Bull. Chem. Soc. Japan* 1974, *47*, 1285-6.
VARIABLES:	PREPARED BY:
One temperature: 25°C	Orest Popovych

EXPERIMENTAL VALUES:

The solubility of $Ru(phen)_3(BPh_4)_2$ in ethanol was reported as:

1.30×10^{-5} mol dm^{-3}.

AUXILIARY INFORMATION

METHOD/APPARATUS/PROCEDURE:	SOURCE AND PURITY OF MATERIALS:
Spectrophotometry. For details see compilation for $Ru(phen)_3(BPh_4)_2$ in acetone based on the same lit. reference.	See compilation for $Ru(phen)_3(BPh_4)_2$ in acetone based on the same lit. reference.
	ESTIMATED ERROR:
	Nothing specified.
	REFERENCES:

COMPONENTS:	ORIGINAL MEASUREMENTS:
(1) Tris(o-phenanthroline)ruthenium (II) tetraphenylborate (1-); $RuC_{84}H_{64}B_2N_6$; (2) Ethyl ethanoate (ethyl acetate); $C_4H_8O_2$; [141-78-6]	Takamatsu, T. *Bull. Chem. Soc. Japan* 1974, *47*, 1285-6.
VARIABLES:	PREPARED BY:
One temperature: 25°C	Orest Popovych

EXPERIMENTAL VALUES:

The solubility of $Ru(phen)_3(BPh_4)_2$ in ethyl acetate was reported as:

5.41×10^{-7} mol dm^{-3}.

AUXILIARY INFORMATION

METHOD/APPARATUS/PROCEDURE:	SOURCE AND PURITY OF MATERIALS:
Spectrophotometry. For details see compilation for $Ru(phen)_3(BPh_4)_2$ in acetone based on the same lit. reference.	See compilation for $Ru(phen)_3(BPh_4)_2$ in acetone based on the same lit. reference.
	ESTIMATED ERROR:
	Nothing specified.
	REFERENCES:

COMPONENTS:	ORIGINAL MEASUREMENTS:
(1) Tris(o-phenanthroline)ruthenium (II) tetraphenylborate (1-); $RuC_{84}H_{64}B_2N_6$; (2) Isopropyl alcohol; C_3H_8O; [67-63-0]	Takamatsu, T. *Bull. Chem. Soc. Japan* 1974, *47*, 1285-6.

VARIABLES:	PREPARED BY:
One temperature: 25°C	Orest Popovych

EXPERIMENTAL VALUES:

The solubility of $Ru(phen)_3(BPh_4)_2$ in isopropyl alcohol was reported as:

$$5.41 \times 10^{-6} \text{ mol dm}^{-3}.$$

AUXILIARY INFORMATION

METHOD/APPARATUS/PROCEDURE:	SOURCE AND PURITY OF MATERIALS:
Spectrophotometry. For details see compilation for $Ru(phen)_3(BPh_4)_2$ in acetone based on the same lit. reference.	See compilation for $Ru(phen)_3(BPh_4)_2$ in acetone based on the same lit. reference.
	ESTIMATED ERROR:
	Nothing specified.
	REFERENCES:

COMPONENTS:	ORIGINAL MEASUREMENTS:
(1) Tris(o-phenanthroline)ruthenium (II) tetraphenylborate (1-); $RuC_{84}H_{64}B_2N_6$; (2) Isopropyl ether; $C_6H_{14}O$; [108-20-3]	Takamatsu, T. *Bull. Chem. Soc. Japan* 1974, *47*, 1285-6.
VARIABLES:	PREPARED BY:
One temperature: 25°C	Orest Popovych

EXPERIMENTAL VALUES:

The solubility of $Ru(phen)_3(BPh_4)_2$ in isopropyl ether was reported as:

$$4.32 \times 10^{-7} \text{ mol dm}^{-3}.$$

AUXILIARY INFORMATION

METHOD/APPARATUS/PROCEDURE:	SOURCE AND PURITY OF MATERIALS:
Spectrophotometry. For details see compilation for $Ru(phen)_3(BPh_4)_2$ in acetone based on the same lit. reference.	See compilation for $Ru(phen)_3(BPh_4)_2$ in acetone based on the same lit. reference.
	ESTIMATED ERROR:
	Nothing specified.
	REFERENCES:

COMPONENTS:

(1) Tris(o-phenanthroline)ruthenium (II) tetraphenylborate (1-); $RuC_{84}H_{64}B_2N_6$;

(2) Methanol; CH_4O; [67-56-1]

ORIGINAL MEASUREMENTS:

Takamatsu, T. *Bull. Chem. Soc. Japan* 1974, *47*, 1285-6.

VARIABLES:

One temperature: 25°C

PREPARED BY:

Orest Popovych

EXPERIMENTAL VALUES:

The solubility of $Ru(phen)_3(BPh_4)_2$ in methanol was reported as:

3.84×10^{-5} mol dm^{-3}.

AUXILIARY INFORMATION

METHOD/APPARATUS/PROCEDURE:

Spectrophotometry. For details see compilation for $Ru(phen)_3(BPh_4)_2$ in acetone based on the same lit. reference.

SOURCE AND PURITY OF MATERIALS:

See compilation for $Ru(phen)_3(BPh_4)_2$ in acetone based on the same lit. reference.

ESTIMATED ERROR:

Nothing specified.

REFERENCES:

COMPONENTS:

(1) Tris(o-phenanthroline)ruthenium (II) tetraphenylborate (1-); $RuC_{84}H_{64}B_2N_6$;

(2) 3-Methyl-1-butanol (isoamyl alcohol); $C_5H_{12}O$; [123-51-3]

ORIGINAL MEASUREMENTS:

Takamatsu, T. *Bull. Chem. Soc. Japan* 1974, *47*, 1285-6.

VARIABLES:

One temperature: 25°C

PREPARED BY:

Orest Popovych

EXPERIMENTAL VALUES:

The solubility of $Ru(phen)_3(BPh_4)_2$ in isoamyl alcohol was reported as:

$$3.78 \times 10^{-6} \text{ mol dm}^{-3}.$$

AUXILIARY INFORMATION

METHOD/APPARATUS/PROCEDURE:

Spectrophotometry. For details see compilation for $Ru(phen)_3(BPh_4)_2$ in acetone based on the same lit. reference.

SOURCE AND PURITY OF MATERIALS:

See compilation for $Ru(phen)_3(BPh_4)_2$ in acetone based on the same lit. reference.

ESTIMATED ERROR:

Nothing specified.

REFERENCES:

COMPONENTS:

(1) Tris(o-phenanthroline)ruthenium (II) tetraphenylborate (1-); $RuC_{84}H_{64}B_2N_6$;

(2) γ-Methyl-butyl ethanoate (iso-amyl acetate); $C_7H_{14}O_2$; [123-92-2]

ORIGINAL MEASUREMENTS:

Takamatsu, T. *Bull. Chem. Soc. Japan* 1974, *47*, 1285-6.

VARIABLES:

One temperature: 25°C

PREPARED BY:

Orest Popovych

EXPERIMENTAL VALUES:

The solubility of $Ru(phen)_3(BPh_4)_2$ in isoamyl acetate was reported as:

$$\sim 10^{-7} \text{ mol dm}^{-3}.$$

AUXILIARY INFORMATION

METHOD/APPARATUS/PROCEDURE:

Spectrophotometry. For details see compilation for $Ru(phen)_3(BPh_4)_2$ in acetone based on the same lit. reference.

SOURCE AND PURITY OF MATERIALS:

See compilation for $Ru(phen)_3(BPh_4)_2$ in acetone based on the same lit. reference.

ESTIMATED ERROR:

Nothing specified.

REFERENCES:

COMPONENTS:

(1) Tris(o-phenanthroline)ruthenium (II) tetraphenylborate (1-); $RuC_{84}H_{64}B_2N_6$;

(2) β-Methyl-propyl ethanoate (isobutyl acetate); $C_6H_{12}O_2$; [110-19-0]

ORIGINAL MEASUREMENTS:

Takamatsu, T. *Bull. Chem. Soc. Japan* 1974, *47*, 1285-6.

VARIABLES:

One temperature: 25°C

PREPARED BY:

Orest Popovych

EXPERIMENTAL VALUES:

The solubility of $Ru(phen)_3(BPh_4)_2$ in isobutyl acetate was reported as:

$$1.08 \times 10^{-7} \text{ mol dm}^{-3}.$$

AUXILIARY INFORMATION

METHOD/APPARATUS/PROCEDURE:

Spectrophotometry. For details see compilation for $Ru(phen)_3(BPh_4)_2$ in acetone based on the same lit. reference.

SOURCE AND PURITY OF MATERIALS:

See compilation for $Ru(phen)_3(BPh_4)_2$ in acetone based on the same lit. reference.

ESTIMATED ERROR:

Nothing specified.

REFERENCES:

COMPONENTS:	ORIGINAL MEASUREMENTS:
(1) Tris(o-phenanthroline)ruthenium (II) tetraphenylborate (1-); $RuC_{84}H_{64}B_2N_6$; (2) 1-Nitropropane; $C_3H_7NO_2$; [25322-01-4]	Takamatsu, T. *Bull. Chem. Soc. Japan* 1974, *47*, 1285-6.
VARIABLES:	PREPARED BY:
One temperature: 25°C	Orest Popovych

EXPERIMENTAL VALUES:

The solubility of $Ru(phen)_3(BPh_4)_2$ in 1-nitropropane was reported as:

$$4.22 \times 10^{-3} \text{ mol dm}^{-3}.$$

AUXILIARY INFORMATION

METHOD/APPARATUS/PROCEDURE:	SOURCE AND PURITY OF MATERIALS:
Spectrophotometry. For details see compilation for $Ru(phen)_3(BPh_4)_2$ in acetone based on the same lit. reference.	See compilations for $Ru(phen)_3(BPh_4)_2$ in acetone based on the same lit. reference.
	ESTIMATED ERROR:
	Nothing specified.
	REFERENCES:

COMPONENTS:

(1) Tris(o-phenanthroline)ruthenium (II) tetraphenylborate (1-); $RuC_{84}H_{64}B_2N_6$;

(2) 2-Propanone (acetone); C_3H_6O; [67-64-1]

ORIGINAL MEASUREMENTS:

Takamatsu, T. *Bull. Chem. Soc. Japan* 1974, *47*, 1285-6.

VARIABLES:

One temperature: 25°C

PREPARED BY:

Orest Popovych

EXPERIMENTAL VALUES:

The solubility of $Ru(phen)_3(BPh_4)_2$ in acetone was reported as:

2.02×10^{-3} mol dm^{-3}.

AUXILIARY INFORMATION

METHOD/APPARATUS/PROCEDURE:

The suspensions were shaken in a thermostat for 7-8 days and analyzed for $Ru(phen)_3^{++}$ spectrophotometrically at 448 nm using a previously determined value for the molar absorptivity of 18,500 dm^3 cm^{-1} mol^{-1} (1). Method of temperature control not specified.

SOURCE AND PURITY OF MATERIALS:

The solvent was repeatedly distilled using a 1-m column packed with helical steel wire. The salt was prepared by reacting 200 mg of $RuCl_3$ H_2O, 526 mg of o-phenanthroline and 218 mg of hydoxylammonium sulfate in 50 cm^3 of 1:1 ethanol-water and the pH adjusted to 6 with satd. $Ba(OH)_2$. After refluxing for 2 days at 100°C, the $BaSO_4$ was filtered out, and the excess o-phenanthroline was removed by washing the solution several times with $CHCl_3$ and then with n-hexane. The resulting solution was dried in vacuo, the residue

Continued....

ESTIMATED ERROR:

Nothing specified.

REFERENCES:

Takamatsu, T. *Bull. Chem. Soc. Japan* 1974, *47*, 118.

COMPONENTS:

(1) Tris(o-phenanthroline)ruthenium (II) tetraphenylborate (1-); $RuC_{84}H_{64}B_2N_6$;

(2) 2-Propanone (acetone); C_3H_6O; [67-64-1]

ORIGINAL MEASUREMENTS:

Takamatsu, T. *Bull. Chem. Soc. Japan* 1974, *47*, 1285-6.

VARIABLES:

PREPARED BY:

COMMENTS AND/OR ADDITIONAL DATA

EXPERIMENTAL VALUES:

AUXILIARY INFORMATION

METHOD/APPARATUS/PROCEDURE:

SOURCE AND PURITY OF MATERIALS:...continued

was dissolved in a minimum of acetonitrile and remaining precipitate filtered out. $Ru(phen)_3Cl$ was crystallized by adding $CHCl_3$ to the filtrate. $Ru(phen)_3(BPh_4)_2$ was prepared by adding a slight excess of $NaBPh_4$ to the chloride. The precipitate was dried for 2 days in vacuo. Elemental analysis: Calcd: C- 78.81%; H- 5.03%; N- 6.56%. Found: C- 78.78; H- 4.81; N- 6.60.

ESTIMATED ERROR:

REFERENCES:

COMPONENTS:

(1) Tris(o-phenanthroline)ruthenium (II) tetraphenylborate (1-); $RuC_{84}H_{64}B_2N_6$;

(2) n-Propyl acetate; $C_5H_{10}O_2$; [109-60-4]

ORIGINAL MEASUREMENTS:

Takamatsu, T. *Bull. Chem. Soc. Japan* 1974, *47*, 1285-6.

VARIABLES:

One temperature: 25°C

PREPARED BY:

Orest Popovych

EXPERIMENTAL VALUES:

The solubility of $Ru(phen)_3(BPh_4)_2$ in n-propyl acetate was reported as:

$$1.62 \times 10^{-6} \text{ mol dm}^{-3}.$$

AUXILIARY INFORMATION

METHOD/APPARATUS/PROCEDURE:

Spectrophotometry. For details see compilation for $Ru(phen)_3(BPh_4)_2$ in acetone based on the same lit. reference.

SOURCE AND PURITY OF MATERIALS:

See compilation for $Ru(phen)_3(BPh_4)_2$ in acetone based on the same lit. reference.

ESTIMATED ERROR:

Nothing specified.

REFERENCES:

COMPONENTS:

(1) Tris(o-phenanthroline)ruthenium (II) tetraphenylborate (1-); $RuC_{84}H_{64}B_2N_6$;

(2) Tetramethylene oxide (tetrahydrofuran); C_4H_8O; [26249-20-7]

ORIGINAL MEASUREMENTS:

Takamatsu, T. *Bull. Chem. Soc. Japan* 1974, *47*, 1285-6.

VARIABLES:

One temperature: 25°C

PREPARED BY:

Orest Popovych

EXPERIMENTAL VALUES:

The solubility of $Ru(phen)_3(BPh_4)_2$ in tetrahydrofuran was reported as:

$$6.03 \times 10^{-4} \text{ mol dm}^{-3}.$$

AUXILIARY INFORMATION

METHOD/APPARATUS/PROCEDURE:

Spectrophotometry. For details see compilation for $Ru(phen)_3(BPh_4)_2$ in acetone based on the same lit. reference.

SOURCE AND PURITY OF MATERIALS:

See compilation for $Ru(phen)_3(BPh_4)_2$ in acetone based on the same lit. reference.

ESTIMATED ERROR:

Nothing specified.

REFERENCES:

COMPONENTS:	ORIGINAL MEASUREMENTS:
(1) Tris(o-phenanthroline)ruthenium (II) tetraphenylborate (1-); $RuC_{84}H_{64}B_2N_6$; (2) 2,2,4,4-tetramethyl-3-pentanone (diisobutyl ketone); $C_9H_{18}O$; [815-24-7]	Takamatsu, T. *Bull. Chem. Soc. Japan* 1974, *47*, 1285-6.
VARIABLES:	PREPARED BY:
One temperature: 25°C	Orest Popovych

EXPERIMENTAL VALUES:

The solubility of $Ru(phen)_3(BPh_4)_2$ in diisobutyl ketone was reported as:

$$1.19 \times 10^{-5} \text{ mol dm}^{-3}.$$

AUXILIARY INFORMATION

METHOD/APPARATUS/PROCEDURE:

Spectrophotometry. For details see compilation for $Ru(phen)_3(BPh_4)_2$ in acetone based on the same lit. reference.

SOURCE AND PURITY OF MATERIALS:

See compilation for $Ru(phen)_3(BPh_4)_2$ in acetone based on the same lit. reference.

ESTIMATED ERROR:

Nothing specified.

REFERENCES:

COMPONENTS:

(1) Silver tetraphenylborate (1-); $AgC_{24}H_{20}B$; [14637-35-5]

(2) Water; H_2O; [7732-18-5]

EVALUATOR:

Orest Popovych, Department of Chemistry, The City University of New York, Brooklyn College, Brooklyn, N. Y. 11210, U. S. A.
September 1979

CRITICAL EVALUATION:

The solubility of silver tetraphenylborate ($AgBPh_4$) in water proved to be exceptionally elusive. It is too low for reliable direct determination by UV-spectrophotometry, since the molar absorptivity of the BPh_4^- ion in water at the characteristic 266-nm peak is only 3.25 x 10^3 dm^3 $(cm\ mol)^{-1}$, while the reported solubility generally ranges between 10^{-7} and 10^{-9} mol dm^{-3}. Furthermore, the reduction of the silver ion in $AgBPh_4$ suspensions by light and possibly by organic impurities creates additional difficulties regardless of the analytical method employed.

Historically, the first to report a solubility product for $AgBPh_4$ in water was Havíř (1). Noting that direct potentionmetry with a silver electrode in saturated $AgBPh_4$ solutions led to irreproducible results, he resorted to potentiometric determinations in the presence of ∿0.01 mol dm^{-3} $NaBPh_4$, which at 293 K yielded a K_{s0} = 4.0 x 10^{-14} (all solubility products in this evalutaion have units of mol^2 dm^{-6}). Popovych (2) repeated Havíř's work at several concentrations of $NaBPh_4$ at 298 K, obtaining a K°_{s0} = 4.07 x 10^{-14}. (The pK_{s0} values from the work of both Havíř (1) and Popovych (2) are 13.4). Alexander et al. (3) reported a pK_{s0} = 11.1 at 298 K and at an ionic strength of 0.01 mol dm^{-3} on the basis of a potentiometric titration of a tetraphenylborate solution with silver ion.

Kolthoff and Chantooni (4) questioned all of the above results on the grounds that the values for the transfer activity coefficients (medium effects) for the transfer of silver and tetraphenylborate ions from water to methanol required a pK°_{s0} value of $AgBPh_4$ in water of the order of 17.5. They repeated Popovych's (2) experiments, obtaining a pK°_{s0} value of 14.3 ± 0.2, from which they concluded that the silver-silver tetraphenylborate electrode behaves abnormally in aqueous solutions, probably because it is not wetted by water (4). Instead, Kolthoff and Chantooni (4) determined the solubility product of $AgBPh_4$ in water from the chemical-exchange experiments in which a solution of $NaBPh_4$ was equilibrated with solid AgI and a solution of NaI was equilibrated with solid $AgBPh_4$. On this basis they reported a pK°_{s0} = 17.2 ± 0.2 at 298K. The latter is probably the best value available to date, but in view of the fact that it has not yet been corroborated by another method or by work from another laboratory that value must be described for now as tentative.

Since the activity correction in this case is certainly negligible, the solubility of $AgBPh_4$ in water can be estimated as $(K^\circ_{s0})^{1/2}$, which leads to the value of Solubility = (2.5 ± 0.3) x 10^{-9} mol dm^{-3} at 298 K. It is, of course, also tentative. In view of this, the solubility of 7.6 x 10^{-6} mol dm^{-3} reported by McClure and Rechnitz (5) for a THAM buffer solution is much too high.

REFERENCES:

1. Havíř, J. *Collect. Czech. Chem. Commun.* 1959, *24*, 1955.
2. Popovych, O. *Anal. Chem.* 1966, *38*, 558.
3. Alexander, R.; Ko, E. C. F.; Mac, Y. C.; Parker, A. J. *J. Am. Chem. Soc.* 1967, *89*, 3703.
4. Kolthoff, I. M.; Chantooni, M. K., Jr. *Anal. Chem.* 1972, *44*, 194.
5. McClure, J. E.; Rechnitz, G. A. *Anal. Chem.* 1966, *38*, 136.

COMPONENTS:

(1) Silver tetraphenylborate (1-); $AgC_{24}H_{20}B$; [14637-35-5]
(2) Sodium tetraphenylborate; $NaC_{24}H_{20}B$; [143-66-8]
(3) Water; H_2O; [7732-18-5]

ORIGINAL MEASUREMENTS:

Havíř, J. *Collect. Czech. Chem. Commun.* 1959, *24*, 1955-9.

VARIABLES:

One temperature: 20°C

PREPARED BY:

Orest Popovych

EXPERIMENTAL VALUES:

The solubility product of $AgBPh_4$ in water determined in the presence of ~0.01 mol dm^{-3} $NaBPh_4$ was reported as:

$$K_{s0}^{\circ} = 4.0 \times 10^{-14} \text{ mol}^2 \text{ dm}^{-6}.$$

Assuming a Nernstein response in the emf cell specified below, the author obtained $a_{Ag} = 4.5 \times 10^{-12}$ mol dm^{-3} and presumably multipled that value by the BPh_4^- activity though no value for the activity coefficient was mentioned specifically.

AUXILIARY INFORMATION

METHOD/APPARATUS/PROCEDURE:

Fifty cm^3 of a 0.01 mol dm^{-3} solution of $NaBPh_4$ were reacted with 2 drops of 0.01 mol dm^{-3} $AgNO_3$ and the resulting suspension shaken in the dark for 4 hrs. A silver electrode dipping in the above suspension was connected via an Agar bridge to another silver electrode in a 0.01 mol dm^{-3} solution of $AgNO_3$. The resulting emf was 0.545 V.

SOURCE AND PURITY OF MATERIALS:

$NaBPh_4$ was from the Heyl & Co. (Berlin).

ESTIMATED ERROR:

Not specified.
Temperature control: ±0.5°C

REFERENCES:

COMPONENTS:

(1) Silver tetraphenylborate (1-); $AgC_{24}H_{20}B$; [14637-35-5]

(2) Sodium tetraphenylborate; $NaC_{24}H_{20}B$; [143-66-8]

(3) Water; H_2O; [7732-18-5]

ORIGINAL MEASUREMENTS:

Popovych, O. *Anal. Chem.* 1966, *38*, 558-63.

VARIABLES:

One temperature: 25.00°C

PREPARED BY:

Orest Popovych

EXPERIMENTAL VALUES:

The author reports the solubility product of $AgBPh_4$ in water as:

$$K_{s0}^{\circ} = (4.07 \pm 0.50) \times 10^{-14} \text{ mol}^2 \text{ dm}^{-6}.$$

AUXILIARY INFORMATION

METHOD/APPARATUS/PROCEDURE:

Potentiometric determination of silver-ion activity in suspensions of $AgBPh_4$ containing BPh_4^- concentrations in the range of 10^{-4}-10^{-2} mol dm^{-3} added as the sodium salt. The potential difference was measured between two silver electrodes, one of which was immersed in the above suspensions and the other, in 10^{-2} mol dm^{-3} $AgNO_3$ solution. No bridge solution was specified. Saturation was achieved by shaking for at least 2 weeks on a Burrell wrist-action shaker in water-jacketed flasks.

SOURCE AND PURITY OF MATERIALS:

$AgBPh_4$ was prepared by metathesis of exactly stoichiometric amounts of aqueous $AgNO_3$ and purified $KBPh_4$ dissolved in a minimum of acetone. The precipitate was washed thoroughly by decantation.

ESTIMATED ERROR:

Relative precision of ±12% in the determination of a_{Ag}.
Temperature: ±0.01°C.

REFERENCES:

COMPONENTS:
(1) Silver tetraphenylborate (1-); $AgC_{24}H_{20}B$; [14637-35-5]
(2) Sodium nitrate; $NaNO_3$; [7631-99-4]
(3) Sodium tetraphenylborate; $NaC_{24}H_{20}B$; [143-66-8]
(4) Water; H_2O; [7732-18-5]

ORIGINAL MEASUREMENTS:

Alexander, R.; Ko, E. C. F.; Mac, Y. C.; Parker, A. J. *J. Am. Chem. Soc.* 1967, *89*, 3703-12.

VARIABLES:

One temperature: 25°C

PREPARED BY:

Orest Popovych

EXPERIMENTAL VALUES:

The solubility product of $AgBPh_4$ in water was reported as a product of ionic concentrations determined at ionic strengths in the range of 0.01-0.005 mol dm^{-3} maintained by $NaBPh_4$:

$$pK_{s0} = 11.1 \ (K_{s0} \text{ units are mol}^2 \text{ dm}^{-6}).$$

AUXILIARY INFORMATION

METHOD/APPARATUS/PROCEDURE:

Potentiometric titration of 0.01 mol dm^{-3} solution of $NaBPh_4$ with $AgNO_3$ using a silver indicator electrode and a silver-silver nitrate reference cell connected by a bridge of NEt_4 picrate. A radiometer pH meter, Type PHM22r was used. Cell was thermostatted, but limits of temperature control were not specified.

SOURCE AND PURITY OF MATERIALS:

The salts were of Analar grade and were used as received.

ESTIMATED ERROR:

None specified. A precision of ±0.1 pK units is assumed by the compiler.

REFERENCES:

COMPONENTS:
(1) Silver tetraphenylborate (1-); $AgC_{24}H_{20}B$; [14627-35-5]
(2) Sodium tetraphenylborate; $NaC_{24}H_{20}B$; [143-66-8]
(3) Sodium iodide; NaI; [7681-82-5]
(4) Water; H_2O; [7732-18-5]

ORIGINAL MEASUREMENTS:
Kolthoff, I. M.; Chantooni, M. K., Jr. *Anal. Chem.* 1972, *44*, 194-5.

VARIABLES:
One temperature: 25°C

PREPARED BY:
Orest Popovych

EXPERIMENTAL VALUES:

The solubility (ion-activity) product of $AgBPh_4$ in water was reported to be:

$$pK^\circ_{s0} = 17.2 \quad (K^\circ_{s0} \text{ units are } mol^2\ dm^{-6}).$$

The above value was derived from chemical-exchange experiments described by the equilibrium:

$$NaBPh_4 + AgI(s) \rightleftarrows AgBPh_4(s) + NaI$$

from which the ratio: $\dfrac{K^\circ_{s0}(AgBPh_4)}{K^\circ_{s0}(AgI)} = \dfrac{[BPh_4^-]}{[I^-]}$ was measured.

The literature value for $pK^\circ_{s0}(AgI)$ used in the above calculation was 16.0 (1). Complete dissociation was assumed for all electrolytes. Activity coefficients were calculated from the limiting Debye-Hückel law. The relevant experimental data are shown on the next page.

AUXILIARY INFORMATION

METHOD/APPARATUS/PROCEDURE:
A suspension of AgI in a solution of $NaBPh_4$ or a suspension of $AgBPh_4$ in a solution of NaI were shaken under deaerated conditions for 5 days. In the former case, $AgBPh_4$ solid was added as seed. The filtered saturated solutions were analyzed spectrophotometrically for the BPh_4^- in the 260-280-nm range, after subtracting the iodide absorbance, and potentiometrically for the I^-, using a silver electrode. UV spectra were recorded on a Cary Model 15 spectrophotometer. Emf measurements were made with a Corning Model 10 pH meter on the millivolt scale.

SOURCE AND PURITY OF MATERIALS:
Conductivity water was used. NaI was Mallinckrodt AR Grade, recrystallized from water. $NaBPh_4$ was Aldrich puriss. product, purified according to a literature method (2). $AgBPh_4$ was prepared by metathesis of $AgNO_3$ and 2% excess of $NaBPh_4$. The product was washed well with methanol, dried in vacuo at 50°C for 3 hrs., all preparation and storage taking place in the dark. AgI was prepared and handled analogously.

ESTIMATED ERROR:
Accuracy ±0.2 pK units (3). A precision of ±0.1 pK units is assumed by the compiler.
Temperature control: ±1°C

REFERENCES:

(1) Buckley, P.; Hartley, H. *Phil. Mag.* 1929, *8*, 320.
(2) Popov, A. I.; Humphrey, R. *J. Am. Chem. Soc.* 1959, *81*, 2043.
(3) Kolthoff, I. M.; Chantooni, M. K., Jr. *J. Phys. Chem.* 1972, *76*, 2024.

COMPONENTS:

(1) Silver tetraphenylborate (1-); $AgC_{24}H_{20}B$; [14627-35-5]
(2) Sodium tetraphenylborate; $NaC_{24}H_{20}B$; [143-66-8]
(3) Sodium iodide; NaI; [7681-82-5]
(4) Water; H_2O; [7732-18-5]

ORIGINAL MEASUREMENTS:

Kolthoff, I. M.; Chantooni, M. K., Jr. *Anal. Chem.* 1972, *44*, 194-5.

VARIABLES:

One temperature: 25°C

PREPARED BY:

Orest Popovych

EXPERIMENTAL VALUES: (continued....)

Exchange experiments

Initial C/mol dm^{-3} $NaBPh_4$ or NaI	$[BPh_4^-]$/mol dm^{-3}	$[I^-]$/mol dm^{-3}	$K°_{s0}(AgBPh_4)$/$K°_{s0}(AgI)$	$pK°_{s0}$ ($AgBPh_4$)
	$NaBPh_4$ + AgI(s)			
0.00740	4.63×10^{-4}	6.94×10^{-3}	0.066	17.2
0.00817	5.1×10^{-4}	7.66×10^{-3}	0.067	17.2
0.0155	1.17×10^{-3}	1.43×10^{-2}	0.082	17.1
	NaI + $AgBPh_4$(s)			
0.0107	5.48×10^{-4}	1.01×10^{-2}	0.054	17.3
0.0214	$1.0_7 \times 10^{-3}$	2.03×10^{-2}	0.053	17.3

AUXILIARY INFORMATION

METHOD/APPARATUS/PROCEDURE:

SOURCE AND PURITY OF MATERIALS:

ESTIMATED ERROR:

REFERENCES:

COMPONENTS:	ORIGINAL MEASUREMENTS:
(1) Silver tetraphenylborate (1-); $AgC_{24}H_{20}B$; [14637-35-5] (2) Tris(hydroxymethyl)aminomethane; $C_4H_{11}NO_3$; [77-86-1] (3) Acetic acid; $C_2H_4O_2$: [64-19-7] (4) Water; H_2O; [7732-18-5]	McClure, J. E.; Rechnitz, G. A. *Anal. Chem.* 1966, *38*, 136-9.
VARIABLES:	**PREPARED BY:**
One temperature: 24.8 C	Orest Popovych

EXPERIMENTAL VALUES:

The solubility of silver tetraphenylborate ($AgBPh_4$) in aqueous solution of the tris(hydroxymethyl)aminomethane (THAM) buffer at pH 5.1 was reported to be 7.6×10^{-6} mol dm^{-3}.

AUXILIARY INFORMATION

METHOD/APPARATUS/PROCEDURE:

UV-spectrophotometry according to the procedure of Howick and Pflaum (1). No other details.

SOURCE AND PURITY OF MATERIALS:

Baker reagent-grade $AgNO_3$ was the starting material for the $AgBPh_4$. $Ca(BPh_4)_2$ was prepared from Fisher Scientific reagent-grade $NaBPh_4$ by the procedure of Rechnitz et al. (2), and was standardized by potentiometric titrn with KCl and RbCl. $Ca(BPh_4)_2$ solution in THAM was the source of BPh_4^-. The buffer contained 0.1 mol dm^{-3} THAM and 0.01 mol dm^{-3} acetic acid, adjusted to pH 5.1 with G. F. Smith reagent-grade $HClO_4$.

ESTIMATED ERROR:

Not stated.
Temperature: ±0.3°C

REFERENCES:

1. Howick, L. C.; Pflaum, R. T. *Anal. Chim. Acta* 1958, *19*, 342.
2. Rechnitz, G. A.; Katz, S. A.; Zamochnick, S. B. *Anal. Chem.* 1963, *35*, 1322.

COMPONENTS:

(1) Silver tetraphenylborate (1-); $AgC_{24}H_{20}B$; [14637-35-5]
(2) Isopropyl alcohol; C_3H_8O; [67-63-0]
(3) Toluene; C_7H_8; [108-88-3]
(4) Water; H_2O; [7732-18-5]

ORIGINAL MEASUREMENTS:

Popovych, O. *Anal. Chem.* 1966, *38*, 117-9.

VARIABLES:

One temperature: 25.00°C

PREPARED BY:

Orest Popovych

EXPERIMENTAL VALUES:

The solubility of $AgBPh_4$ in the mixture consisting of 50.0% toluene, 49.5% isopropyl alcohol and 0.5% water by volume known as the ASTM medium* was reported as:

$$C = 5.40 \times 10^{-5}\ \text{mol dm}^{-3}.$$

The corresponding ionic concentration in the saturated solution was reported to be:

$$C\alpha = 2.48 \times 10^{-5}\ \text{mol dm}^{-3}.$$

The molar activity coefficient for the electrolyte in the ASTM medium at 25°C is given by the limiting Debye-Hückel expression as:

$$\log y_{\pm}^{2} = -31.44(C\alpha)^{1/2}.$$

The solubility product for $AgBPh_4$ in the ASTM medium calculated from the above results in the form: $K_{s0}^{\circ} = (C\alpha y_{\pm})^2$ was reported as:

$$K_{s0}^{\circ} = 4.29 \times 10^{-10}\ \text{mol}^2\ \text{dm}^{-6}.$$

The degree of dissociation α was calculated from the Wirth equation (1):

$$\alpha = \frac{1000\ \kappa}{C[\Lambda^{\infty} - S(1000\ \kappa/\Lambda^{\infty})]^{1/2}},$$

...continued

AUXILIARY INFORMATION

METHOD/APPARATUS/PROCEDURE:

Electrolytic conductivity of saturated and diluted solutions in conjunction with equations developed in this article. The measurements were carried out on a Wayne-Kerr Universal Bridge B221 with a platinum cell. Saturated solutions were prepared by shaking for at least 2 weeks on a Burrell wrist-action shaker in water-jacketed flasks.

SOURCE AND PURITY OF MATERIALS:

$AgBPh_4$ was prepared by metathesis of exactly stoichiometric amounts of aqueous $AgNO_3$ and purified $KBPh_4$ dissolved in a minimum of acetone. The precipitate was washed by decantation first with water, then with ASTM solvent. The purification of solvents was described elsewhere (2).

ESTIMATED ERROR:

None specified, but the precision in the solubility was about ±6% (compiler). Temperature: ±0.01°C.

REFERENCES:

(1) Wirth, H. E. *J. Phys. Chem.* 1961, *65*, 1441.

(2) Popovych, O. *J. Phys. Chem.* 1962, *66*, 915.

COMPONENTS:

(1) Silver tetraphenylborate (1-); $AgC_{24}H_{20}B$; [14637-35-5]
(2) Isopropyl alcohol; C_3H_8O; [67-63-0]
(3) Toluene; C_7H_8; [108-88-3]
(4) Water; H_2O; [7732-18-5]

ORIGINAL MEASUREMENTS:

Popovych, O. *Anal. Chem.* 1966, *38*, 117-9.

VARIABLES:

One temperature: 25.00°C

PREPARED BY:

Orest Popovych

EXPERIMENTAL VALUES: (Continued ....)

where S is the Onsager coefficient and κ and Λ^{∞} are the electrolytic and limiting molar conductivities, respectively.

*American Society for Testing Materials specifies this solvent for acid-base measurements on petroleum products. The composition of the ASTM solvent corresponds to 52.4 mass% toluene, 47.0 mass% isopropyl alcohol and 0.6 mass% water.

AUXILIARY INFORMATION

METHOD/APPARATUS/PROCEDURE:

SOURCE AND PURITY OF MATERIALS:

ESTIMATED ERROR:

REFERENCES:

COMPONENTS:

(1) Silver tetraphenylborate (1-); $AgC_{24}H_{20}B$; [14637-35-5]

(2) Acetonitrile; C_2H_3N; [75-05-8]

EVALUATOR:

Orest Popovych, Department of Chemistry, The City University of New York, Brooklyn College, Brooklyn, N. Y. 11210, U. S. A.
September 1979

CRITICAL EVALUATION:

The solubility product of silver tetraphenylborate ($AgBPh_4$) in acetonitrile was reported in three studies, all determinations being at 298 K (1-3). In their first publication, Alexander et al. (1) reported a pK_{s0} = 7.2 (all K_{s0} units are $mol^2\ dm^{-6}$). It was determined by potentiometric titration of BPh_4^- with Ag^+ at a constant ionic strength of 0.01 mol dm^{-3} using a silver electrode. The revised value from the same laboratory was pK_{s0}° = 7.5 (2). It was also derived from analogous potentiometric titrations, but was based on additional measurements and activity corrections from the Davies equation (see compilation). Unfortunately, neither temperature control, nor a method of ascertaining saturation were mentioned in the above two publications. Fortunately, at the low levels of precision involved, fine temperature control is not likely to be critical.

Kolthoff and Chantooni (3) published a pK_{s0}° value of 7.7, using the same potentiometric method as for $AgBPh_4$ in methanol (see compilation). Although one cannot be sure of the specifics, this implies that activity coefficients may have been calculated by the Debye-Hückel equation with ion-size parameters and that a temperature control of ±1°C was probably maintained, since both measures were observed in the methanol study (4). The analysis was carried out on saturated solutions.

Although the pK_{s0}° of 7.7 may be the best available datum to date, its accuracy was estimated by the authors as not better than ±0.2 pK units. In view of this we may feel justified to average their result with the pK_{s0}° of 7.5 reported by Alexander et al. (2), obtaining as the tentative value pK_{s0}° = 7.6.

Assuming as usual an error of 0.1 pK_{s0}° units, we can estimate the solubility as $(K_{s0})^{1/2}$: Solubility = $(1.6 \pm 0.2) \times 10^{-4}$ mol dm^{-3}. This, of course, is an estimate which neglects the unknown activity correction introduced originally in the calculation of K_{s0}°, so that the solubility of $AgBPh_4$ in acetonitrile cannot even be designated as tentative at this time.

REFERENCES:

1. Alexander, R.; Ko, E. C. F.; Mac, Y. C.; Parker, A. J. *J. Am. Chem. Soc.* 1967, *89*, 3703.
2. Alexander, R.; Parker, A. J.; Sharp, J. H.; Waghorne, W. E. *J. Am. Chem. Soc.* 1972, *94*, 1148.
3. Kolthoff, I. M.; Chantooni, M. K., Jr. *J. Phys. Chem.* 1972, *76*, 2024.
4. Kolthoff, I. M.; Chantooni, M. K., Jr. *Anal. Chem.* 1972, *44*, 194.

COMPONENTS:	ORIGINAL MEASUREMENTS:
(1) Silver tetraphenylborate (1-); $AgC_{24}H_{20}B$; [14637-35-5] (2) Sodium nitrate; $NaNO_3$; [7631-99-4] (3) Sodium tetraphenylborate; $NaC_{24}H_{20}B$; [143-66-8] (4) Acetonitrile; C_2H_3N; [75-05-8]	Alexander, R.; Ko, E. C. F.; Mac, Y. C.; Parker, A. J. *J. Am. Chem. Soc.* 1967, *89*, 3703-12.

VARIABLES:	PREPARED BY:
One temperature: 25°C	Orest Popovych

EXPERIMENTAL VALUES:

The solubility product of $AgBPh_4$ in acetonitrile was reported as a product of ionic concentrations determined at ionic strengths in the range of 0.10-0.05 mol dm^{-3}, maintained by $NaBPh_4$:

$$pK_{s0} = 7.2 \ (K_{s0} \text{ units are mol}^2 \text{ dm}^{-6}).$$

AUXILIARY INFORMATION

METHOD/APPARATUS/PROCEDURE:

Potentiometric titration of 0.01 mol dm^{-3} solution of $NaBPh_4$ with $AgNO_3$ using a silver indicator electrode and a silver-silver nitrate reference cell connected by a bridge of NEt_4 picrate. A radiometer pH meter, Type PHM22r was used. Cell was thermostatted, but limits of temperature control were not specified.

SOURCE AND PURITY OF MATERIALS:

Acetonitrile was purified by a literature method (1). The salts were of Analar grade and were used as received.

ESTIMATED ERROR:

None specified. A precision of ±0.1 pK units is assumed by the compiler.

REFERENCES:

(1) Coetzee, J. F.; Cunningham, G. P.; McGuire, D. K.; Padmanabhan, G. R.; *Anal. Chem.* 1962, *34*, 1139.

COMPONENTS:

(1) Silver tetraphenylborate (1-); $AgC_{24}H_{20}B$; [14637-35-5]
(2) Sodium perchlorate; $NaClO_4$; [7601-89-0]
(3) Sodium tetraphenylborate; $NaC_{24}H_{20}B$; [143-66-8]
(4) Acetonitrile; C_2H_3N; [75-05-8]

ORIGINAL MEASUREMENTS:

Alexander, R.; Parker, A. J.; Sharp, J. H.; Waghorne, W. E. *J. Am. Chem. Soc.* 1972, *94*, 1148-58.

VARIABLES:

One temperature: 25°C

PREPARED BY:

Orest Popovych

EXPERIMENTAL VALUES:

The solubility (ion-activity) product of $AgBPh_4$ in acetonitrile was determined in the presence of 0.01-0.005 mol dm^{-3} $NaBPh_4$ and reported as:

$$pK^{\circ}_{s0} = 7.5 \quad (K^{\circ}_{s0} \text{ units are mol}^2 \text{ dm}^{-6}).$$

The mean ionic activity coefficient was calculated from the Davies equation in the form: $\log \gamma_{\pm} = -A\,[(I)^{1/2}/(1 + (I)^{1/2}) - (1/3)\,I]$, where I is the ionic strength in mol dm^{-3} and the value of A used was 1.543 $mol^{-1/2}\ dm^{3/2}$.

AUXILIARY INFORMATION

METHOD/APPARATUS/PROCEDURE:

Potentiometric tetration of 0.01 mol dm^{-3} $NaBPh_4$ with 0.01 mol dm^{-3} $AgClO_4$. The ionic strength at the midpoint of the titration curve was used to calculate the activity coefficient.

SOURCE AND PURITY OF MATERIALS:

Not stated.

ESTIMATED ERROR:

Not specifed. A precision of ±0.1 pK units is assumed by the compiler.

REFERENCES:

COMPONENTS:

(1) Silver tetraphenylborate (1-); $AgC_{24}H_{20}B$; [14637-35-5]
(2) Sodium tetraphenylborate; $NaC_{24}H_{20}B$; [143-66-8]
(3) Acetonitrile; C_2H_3N; [75-05-8]

ORIGINAL MEASUREMENTS:

Kolthoff, I. M.; Chantooni, M. K., Jr. *J. Phys. Chem.* 1972, *76*, 2024-34.

VARIABLES:

One temperature: 25°C

PREPARED BY:

Orest Popovych

EXPERIMENTAL VALUES:

The solubility product of $AgBPh_4$ in acetonitrile was reported as:

$$pK^{\circ}_{s0} = 7.7 \; (K^{\circ}_{s0} \text{ units are in mol}^2 \text{ dm}^{-6}).$$

AUXILIARY INFORMATION

METHOD/APPARATUS/PROCEDURE:

Potentiometric determination of pa_{Ag} in saturated solution in the presence of $NaBPh_4$, as described for methanol (1). It should be noted, however, that the reference cited by the authors (Kolthoff, I. M.; Chantooni, M. K., Jr. *J. Am. Chem. Soc.* 1971, *93*, 7104.) does not pertain to the methanol study. The authors reported that $AgBPh_4$ does not form solvates with acetonitrile. (See compilation for $AgBPh_4$ in methanol based on Ref. (1)).

SOURCE AND PURITY OF MATERIALS:

Literature methods were used to purify acetonitrile (2) and $NaBPh_4$ (Aldrich puriss. grade) (3). The preparation of $AgBPh_4$ was not outlined in this article and was probably accomplished by metathesis of $NaBPh_4$ and $AgNO_3$ as described earlier (1).

ESTIMATED ERROR:

Only an accuracy of ±0.2 pK units was mentioned.
A precision of ±0.1 pK units is assumed by the compiler.

REFERENCES:

(1) Kolthoff, I. M.; Chantooni, M. K. Jr. *Anal. Chem.* 1972, *44*, 194.
(2) Kolthoff, I. M.; Bruckenstein, S.; Chantooni, M. K., Jr. *J. Am. Chem. Soc.* 1961, *83*, 3927.
(3) Popov, A. I.; Humphrey, R. *J. Am. Chem. Soc.* 1959, *81*, 2043.

COMPONENTS:
(1) Silver tetraphenylborate (1-); $AgC_{24}H_{20}B$; [14637-35-5]
(2) Sodium nitrate; $NaNO_3$; [7631-99-4]
(3) Sodium tetraphenylborate; $NaC_{24}H_{20}B$; [143-66-8]
(4) N,N-Dimethylacetamide; C_4H_9NO; [127-19-5]

ORIGINAL MEASUREMENTS:
Alexander, R.; Ko, E. C. F.; Mac, Y. C.; Parker, A. J. *J. Am. Chem. Soc.* 1967, *89*, 3703-12.

VARIABLES:
One temperature: 25°C

PREPARED BY:
Orest Popovych

EXPERIMENTAL VALUES:

The solubility product of $AgBPh_4$ in N,N-dimethylacetamide was reported as a product of ionic concentrations determined at ionic strengths in the range of 0.10-0.05 mol dm^{-3}, maintained by $NaBPh_4$:

$$pK_{s0} = 5.9 \ (K_{s0} \text{ units are mol}^2 \text{ dm}^{-6}).$$

AUXILIARY INFORMATION

METHOD/APPARATUS/PROCEDURE:
Potentiometric titration of 0.01 mol dm^{-3} solution of $NaBPh_4$ with $AgNO_3$ using a silver indicator electrode and a silver-silver nitrate reference cell connected by a bridge of NEt_4 picrate. A radiometer pH meter, Type PHM22r was used. Cell was thermostatted, but limits of temperature control were not specified.

SOURCE AND PURITY OF MATERIALS:
N,N-Dimethylacetamide was dried with Type 4A molecular sieves and fractionated twice under a reduced pressure of dry nitrogen. The salts were of Analar grade and were used as received.

COMMENTS AND/OR ADDITIONAL DATA:
The above solubility product seems to be a combination of Ag^+ activity and BPh_4^- concentration. The validity of the reported value is therefore doubtful.

ESTIMATED ERROR:
Nothing specified. A precision of ±0.1 pK units is assumed by the compiler.

REFERENCES:

COMPONENTS:

(1) Silver tetraphenylborate (1-); $AgC_{24}H_{20}B$; [14637-35-5]

(2) N,N-Dimethylformamide; C_3H_7NO; [68-12-2]

EVALUATOR:

Orest Popovych, Department of Chemistry, The City University of New York, Brooklyn College, Brooklyn, N. Y. 11210, U. S. A.
September 1979

CRITICAL EVALUATION:

The solubility product of silver tetraphenylborate ($AgBPh_4$) in N,N-dimethylformamide at 298 K was published in three literature sources (1-3). Both values reported by Alexander et al. (1, 2) were derived from potentiometric titrations of BPh_4^- with Ag^+ using a silver electrode. In their first study, a concentration solubility product was reported in the form pK_{s0} = 6.7 (1). (All K_{s0} units are $mol^2\ dm^{-6}$). In the follow-up study from the same laboratory, which involved additional determinations and activity corrections via the Davies equation (see compilation), the revised value was pK°_{s0} = 7.1 (2). Unfortunately, neither temperature control, nor a method of ascertaining saturation were mentioned in the above two publications. However, at the low levels of precision involved, a fine temperature control may not be critical.

Kolthoff and Chantooni (3) reported a pK°_{s0} = 7.5 determined by the same potentiometric procedure as for $AgBPh_4$ in methanol (see compilation). If this means that activity coefficients were calculated from the Debye-Hückel equation with ion-size parameters and that temperature was controlled to ±1°C, as was done on both counts in methanol (4), their result may be slightly more reliable than that of Alexander et al. (2). At least the analysis was carried out in definitely saturated solutions, as opposed to the assumed saturation at an electrode in the course of a potentiometric precipitation titration. Thus the pK°_{s0} = 7.5 may be described as the tentative value for $AgBPh_4$ in N,N-dimethylformamide. The corresponding solubility taken as $(K^\circ_{s0})^{1/2}$ = (1.8 ± 0.2) x 10^{-4} mol dm^{-3}, assuming a precision of 0.1 units in pK°_{s0}. Of course, this estimate neglects an unknown activity correction contained in the K°_{s0}, so that it cannot even be described as tentative. For example, if we use the above estimate of the solubility to compute the activity coefficient from the Davies equation, the result is $y_\pm$ = 0.954 and the "corrected" solubility value becomes 1.9 x 10^{-4} mol dm^{-3}.

REFERENCES:

1. Alexander, R.; Ko, E. C. F.; Mac, Y. C.; Parker, A. J. *J. Am. Chem. Soc.* 1967, *89*, 3703.
2. Alexander, R.; Parker, A. J.; Sharp, J. H.; Waghorne, W. E. *J. Am. Chem. Soc.* 1972, *94*, 1148.
3. Kolthoff, I. M.; Chantooni, M. K., Jr. *J. Phys. Chem.* 1972, *76*, 2024.
4. Kolthoff, I. M.; Chantooni, M. K., Jr. *Anal. Chem.* 1972, *44*, 194.

COMPONENTS: (1) Silver tetraphenylborate (1-); $AgC_{24}H_{20}B$; [14637-35-5]
(2) Sodium nitrate; $NaNO_3$; [7631-99-4]
(3) Sodium tetraphenylborate; $NaC_{24}H_{20}B$; [143-66-8]
(4) N,N-Dimethylformamide; C_3H_7NO; [68-12-2]

ORIGINAL MEASUREMENTS:

Alexander, R.; Ko, E. C. F.; Mac, Y. C.; Parker, A. J. *J. Am. Chem. Soc.* 1967, *89*, 3703-12.

VARIABLES:

One temperature: 25°C

PREPARED BY:

Orest Popovych

EXPERIMENTAL VALUES:

The solubility product of $AgBPh_4$ in N,N-dimethylformamide was reported as a product of ionic concentrations determined at ionic strengths in the range of 0.10-0.05 mol dm^{-3}, maintained by $NaBPh_4$:

$$pK_{s0} = 6.7 \ (K_{s0} \text{ units are mol}^2 \text{ dm}^{-6}).$$

AUXILIARY INFORMATION

METHOD/APPARATUS/PROCEDURE:

Potentiometric titration of 0.01 mol dm^{-3} solution of $NaBPh_4$ with $AgNO_3$ using a silver indicator electrode and a solver-silver nitrate reference cell connected by a bridge of NEt_4 picrate. A radiometer pH meter, Type PHM22r was used. Cell was thermostatted, but limits of temperature control were not specified.

SOURCE AND PURITY OF MATERIALS:

N,N-Dimethylformamide was dried with Type 4A molecular sieves and fractionated twice under a reduced pressure of dry nitrogen. The salts were of Analar grade and were used as received.

ESTIMATED ERROR:

None specified. A precision of ±0.1 pK units is assumed by the compiler.

REFERENCES:

COMPONENTS: (1) Silver tetraphenylborate (1-); $AgC_{24}H_{20}B$; [14637-35-5]
(2) Sodium perchlorate; $NaClO_4$; [7601-89-0]
(3) Sodium tetraphenylborate; $NaC_{24}H_{20}B$; [143-66-8]
(4) N,N-Dimethylformamide; C_3H_7NO; [68-12-2]

ORIGINAL MEASUREMENTS:

Alexander, R.; Parker, A. J.; Sharp, J. H.; Waghorne, W. E. *J. Am. Chem. Soc.* 1972, *94*, 1148-58.

VARIABLES:

One temperature: 25°C

PREPARED BY:

Orest Popovych

EXPERIMENTAL VALUES:

The solubility (ion-activity) product of $AgBPh_4$ in N,N-dimethylformamide was determined in the presence of 0.01-0.005 mol dm^{-3} $NaBPh_4$ and reported as:

$$pK^{\circ}_{s0} = 7.1 \ (K^{\circ}_{s0} \text{ units are mol}^2 \text{ dm}^{-6}).$$

The mean ionic activity coefficient was calculated from the Davies equation in the form: $\log \gamma_{\pm} = -A\ [(I)^{1/2}/(1 + (I)^{1/2}) - (1/3)I]$, where $\underline{I}$ is the ionic strength in mol dm^{-3} and the value of $\underline{A}$ used was 1.551 $mol^{-1/2}\ dm^{3/2}$.

AUXILIARY INFORMATION

METHOD/APPARATUS/PROCEDURE:

Potentionmetric titration of 0.01 mol dm^{-3} $NaBPh_4$ with 0.01 mol dm^{-3} $AgClO_4$. The ionic strength at the midpoint of the titration curve was used to calculate the activity coefficient.

SOURCE AND PURITY OF MATERIALS:

Not stated.

ESTIMATED ERROR:

Not specified. A precision of ±0.1 pK units is assumed by the compiler.

REFERENCES:

COMPONENTS:

(1) Silver tetraphenylborate (1-); $AgC_{24}H_{20}B$; [14637-35-5]

(2) Sodium tetraphenylborate; $NaC_{24}H_{20}B$; [143-66-8]

(3) N,N-Dimethylformamide; C_3H_7NO; [68-12-2]

ORIGINAL MEASUREMENTS:

Kolthoff, I. M.; Chantooni, M. K., Jr. *J. Phys. Chem.* 1972, *76*, 2024-34.

VARIABLES:

One temperature: 25°C

PREPARED BY:

Orest Popovych

EXPERIMENTAL VALUES:

The solubility product of $AgBPh_4$ in N,N-dimethylformamide was reported as:

$$pK^{\circ}_{s0} = 7.5 \text{ } (K^{\circ}_{s0} \text{ units are in } mol^2 \ dm^{-6}).$$

AUXILIARY INFORMATION

METHOD/APPARATUS/PROCEDURE:

Potentiometric determination of p_{Ag} on filtered saturated solutions in the presence of $NaBPh_4$ as described for methanol (1). (See compilation for methanol based on Ref. 1) However, the reference cited by the authors (Kolthoff, I. M.; Chantooni, M. K., Jr. *J. Am. Chem. Soc.* 1971, *93*, 7104.) does not pertain to the methanol study. The authors detected no crystal solvates of $AgBPh_4$ with N,N-dimethylformamide.

SOURCE AND PURITY OF MATERIALS:

N,N-Dimethylformamide was purified as described in the literature (2) as was $NaBPh_4$ (Aldrich puriss. grade) (3). The preparation of $AgBPh_4$ was not described, but most likely involved the metathesis of $AgNO_3$ and $NaBPh_4$ as in an earlier study (1).

ESTIMATED ERROR:

Nothing specified, except an accuracy of ±0.2 pK units. A precision of ±0.1 pK units is assumed by the compiler.

REFERENCES:

(1) Kolthoff, I. M.; Chantooni, M. K. Jr. *Anal. Chem.* 1972, *44*, 194.

(2) Kolthoff, I. M.; Chantooni, M. K. Jr.; Smagowski, H. *Anal. Chem.* 1970, *42*, 1622.

(3) Popov, A. I.; Humphrey, R. *J. Am. Chem. Soc.* 1959, *81*, 2043.

COMPONENTS:

(1) Silver tetraphenylborate (1-); $AgC_{24}H_{20}B$; [14637-35-5]

(2) Dimethylsulfoxide; C_2H_6OS; [67-68-5]

EVALUATOR:

Orest Popovych, Department of Chemistry, The City University of New York, Brooklyn College, Brooklyn, N. Y. 11210, U. S. A.
September 1979

CRITICAL EVALUATION:

The solubility product of silver tetraphenylborate ($AgBPh_4$) in dimethylsulfoxide was reported in three literature sources, all determinations being at 298 K (1-3). Using either potentiometry or UV-spectrophotometry, Alexander et al. (1) determined a pK_{s0} = 4.6 from the product of ionic concentrations (all K_{s0} units are $mol^2\ dm^{-6}$). A subsequent report from the same laboratory (2) based upon potentiometric titration of BPh_4^- with Ag^+ using a silver electrode and activity corrections by the Davies equation (see compilation) gave a pK°_{s0} = 5.1. It is difficult to evaluate these results, as neither temperature control, nor a method for ascertaining saturation have been specified.

Kolthoff and Chantooni (3) reported a potentionmetric pK°_{s0} value of 4.7_5 and a conductometric value of 4.7, with an estimated accuracy of ±0.2 pK units. The analysis was carried out on filtered saturated solutions in both cases, but nothing is mentioned about activity corrections or temperature control. In analogous work reported from the same laboratory (4), the temperature control was ±1°C and activity coefficients were calculated from a Debye-Hückel equation with ion-size parameters. Thus, the value pK°_{s0} = 4.7_5 can be considered no better than <u>tentative</u> at this time.

In view of the uncertainty in the activity corrections incorporated in the above pK°_{s0} value, the solubility must be derived from the concentration pK_{s0} determined by Alexander et al. (1). Assuming an error of 0.1 pK units in the value 4.6, we obtain for the <u>solubility (5.0 ± 0.6) x 10^{-3} mol dm^{-3}.</u> (evaluator). This solubility value must be described as highly <u>tentative</u>.

REFERENCES:

1. Alexander, R.; Ko, E. C. F.; Mac, Y. C.; Parker, A. J. *J. Am. Chem. Soc.* 1967, *89*, 3703.
2. Alexander, R.; Parker, A. J.; Sharp, J. H.; Waghorne, W. E. *J. Am. Chem. Soc.* 1972, *94*, 1148.
3. Kolthoff, I. M.; Chantooni, M. K., Jr. *J. Phys. Chem.* 1972, *76*, 2024.
4. Kolthoff, I. M.; Chantooni, M. K., Jr. *Anal. Chem.* 1972, *44*, 194.

COMPONENTS:

(1) Silver tetraphenylborate (1-); $AgC_{24}H_{20}B$; [14637-35-5]

(2) Dimethylsulfoxide; C_2H_6OS; [67-68-5]

ORIGINAL MEASUREMENTS:

Alexander, R.; Ko, E. C. F.; Mac, Y. C.; Parker, A. J. *J. Am. Chem. Soc.* 1967, *89*, 3703-12.

VARIABLES:

One temperature: 25°C

PREPARED BY:

Orest Popovych

EXPERIMENTAL VALUES:

The solubility product of $AgBPh_4$ in dimethylsulfoxide was reported as a product of ionic concentration at the ionic strength corresponding to the solubility:

$$pK_{s0} = 4.6 \ (K_{s0} \text{ units are mol}^2 \text{ dm}^{-6}).$$

AUXILIARY INFORMATION

METHOD/APPARATUS/PROCEDURE:

Not specified unambiguously: it could have been either potentiometric titration with KI or UV-spectrophotometry on a Unicam SP500 spectrophotometer. Saturated solutions were prepared by shaking for 24 hours at 35°C and then for a further 24 hours at 25°C.

SOURCE AND PURITY OF MATERIALS:

Dimethylsulfoxide (Crown Zellerbach Corp.) was dried with Type 4A molecular sieves and fractionated twice under a reduced pressure of dry nitrogen. Analar grade salts were used as received.

ESTIMATED ERROR:

None specified. A precision of ±0.1 pK units is assumed by the compiler.

REFERENCES:

COMPONENTS:	ORIGINAL MEASUREMENTS:
(1) Silver tetraphenylborate (1-); $AgC_{24}H_{20}B$; [14637-35-5] (2) Sodium perchlorate; $NaClO_4$; [7601-89-0] (3) Sodium tetraphenylborate; $NaC_{24}H_{20}B$; [143-66-8] (4) Dimethylsulfoxide; C_2H_6OS; [67-68-5]	Alexander, R.; Parker, A. J.; Sharp, J. H.; Waghorne, W. E. *J. Am. Chem. Soc.* 1972, *94*, 1148-58.
VARIABLES:	PREPARED BY:
One temperature: 25°C	Orest Popovych

EXPERIMENTAL VALUES:

The solubility (ion-activity) product of $AgBPh_4$ in dimethylsulfoxide was determined in the presence of 0.01-0.005 mol dm^{-3} $NaBPh_4$ and reported as:

$$pK^{\circ}_{s0} = 5.1 \text{ } (K^{\circ}_{s0} \text{ units are mol}^2 \text{ dm}^{-6}).$$

The mean ionic activity coefficient was calculated from the Davies equation in the form: $\log \gamma_{\pm} = -A\ [(I)^{1/2}/(1 + (I)^{1/2}) - (1/3)\ I]$, where $\underline{I}$ is the ionic strength in mol dm^{-3} and the value of $\underline{A}$ used was 1.115 $mol^{-1/2}\ dm^{3/2}$.

AUXILIARY INFORMATION

METHOD/APPARATUS/PROCEDURE:

Potentiometric tetration of 0.01 mol dm^{-3} $NaBPh_4$ with 0.01 mol dm^{-3} $AgClO_4$. The ionic strength at the midpoint of the titration curve was used to calculate the activity coefficient.

SOURCE AND PURITY OF MATERIALS:

Not stated.

ESTIMATED ERROR:

Nothing specified. A precision of ±0.1 pK units is assumed by the compiler.

REFERENCES:

COMPONENTS:

(1) Silver tetraphenylborate (1-); $AgC_{24}H_{20}B$; [14637-35-5]

(2) Sodium tetraphenylborate $NaC_{24}H_{20}B$; [143-66-8]

(3) Dimethylsulfoxide; C_2H_6OS; [67-68-5]

ORIGINAL MEASUREMENTS:

Kolthoff, I. M.; Chantooni, M. K., Jr. *J. Phys. Chem.* 1972, *76*, 2024-34.

VARIABLES:

One temperature: 25°C

PREPARED BY:

Orest Popovych

EXPERIMENTAL VALUES:

The solubility product of $AgBPh_4$ in dimethylsulfoxide was reported as:

$$pK^\circ_{s0} = 4.7_5 \ (K^\circ_{s0} \text{ units are mol}^2 \text{ dm}^{-6}).$$

AUXILIARY INFORMATION

METHOD/APPARATUS/PROCEDURE:

Potentiometric determination of pa_{Ag} in a filtered saturated solution in the presence of $NaBPh_4$, as described for methanol (1). However, the reference cited by the authors (Kolthoff, I. M.; Chantooni, M. K., Jr. *J. Am. Chem. Soc.* 1971, *93*, 7104.) does not pertain to the methanol study.

SOURCE AND PURITY OF MATERIALS:

Dimethylsulfoxide was purified by a literature method (2), as was $NaBPh_4$ (Aldrich puriss. grade) (3). The preparation of $AgBPh_4$ was not described, but most likely involved the metathesis of $AgNO_3$ and $NaBPh_4$ as in an earlier study (1).

ESTIMATED ERROR:

Nothing specified, except an accuracy of ±0.2 pK units. A precision of ±0.1 pK units is assumed by the compiler.

REFERENCES:

(1) Kolthoff, I. M.; Chantooni, M. K. Jr. *Anal. Chem.* 1972, *44*, 194.
(2) Kolthoff, I. M.; Reddy, T. B. *Inorg. Chem.* 1962, *1*, 189.
(3) Popov, A. I.; Humphrey, R. J. *J. Am. Chem. Soc.* 1959, *81*, 2043.

COMPONENTS:

(1) Silver tetraphenylborate (1-); $AgC_{24}H_{20}B$; [14637-35-5]

(2) Dimethylsulfoxide; C_2H_6OS; [67-68-5]

ORIGINAL MEASUREMENTS:

Kolthoff, I. M.; Chantooni, M. K., Jr. *J. Phys. Chem.* 1972, *76*, 2024-34.

VARIABLES:

One temperature: 25°C

PREPARED BY:

Orest Popovych

EXPERIMENTAL VALUES:

The solubility product of $AgBPh_4$ in dimethylsulfoxide was reported as:

$$pK^{\circ}_{s0} = 4.7$$

AUXILIARY INFORMATION

METHOD/APPARATUS/PROCEDURE:

Conductometric determination on filtered saturated solutions using apparatus previously described (1). Presumably the solubility C was calculated from the relationship $C = 10^3\kappa/\Lambda^{\infty}$, but this was not explained and the Λ^{∞} value was not specified. The reported value of κ was 1.21×10^{-4} ohm^{-1} cm^{-1}.

SOURCE AND PURITY OF MATERIALS:

Dimethylsulfoxide was purified by a literature method (2), as was $NaBPh_4$ (Aldrich puriss. grade) (3). The preparation of $AgBPh_4$ was not described, but probably involved the metathesis of $AgNO_3$ and $NaBPh_4$, as in a previous study (4).

ESTIMATED ERROR:

Precision of conductance data: ±2%.

REFERENCES: (1) Kolthoff, I. M.; Bruckenstein, S.; Chantooni, M. K., Jr. *J. Am. Chem. Soc.* 1961, *83*, 3927.
(2) Kolthoff, I. M.; Reddy, T. B. *Inorg. Chem.* 1962, *1*, 189.
(3) Popov, A. I.; Humphrey, R. *J. Am. Chem. Soc.* 1959, *81*, 2043.
(4) Kolthoff, I. M.; Chantooni, M. K., Jr. *Anal. Chem.* 1972, *44*, 194.

COMPONENTS:

(1) Silver tetraphenylborate (1-); $AgC_{24}H_{20}B$; [14637-35-5]

(2) Sodium perchlorate; $NaClO_4$; [7601-89-0]

(3) Sodium tetraphenylborate; $NaC_{24}H_{20}B$; [143-66-8]

(4) Ethanol; C_2H_6O; [64-17-5]

ORIGINAL MEASUREMENTS:

Alexander, R.; Parker, A. J.; Sharp, J. H.; Waghorne, W. E. *J. Am. Chem. Soc.* 1972, *94*, 1148-58.

VARIABLES:

One temperature: 25°C

PREPARED BY:

Orest Popovych

EXPERIMENTAL VALUES:

The solubility (ion-activity) product of $AgBPh_4$ in ethanol was determined in the presence of 0.01-0.005 mol dm^{-3} $NaBPh_4$ and reported as:

$$pK^\circ_{s0} = 14.4 \quad (K^\circ_{s0} \text{ units are mol}^2 \text{ dm}^{-6}).$$

The mean ionic activity coefficient was calculated from the Davies equation in the form: $\log \gamma_{\pm} = -A\,[(I)^{1/2}/(1 + (I)^{1/2}) - (1/3)I]$, where $\underline{I}$ is the ionic strength in mol dm^{-3} and the value of $\underline{A}$ used was 2.956 $mol^{-1/2}$ $dm^{3/2}$. The solubility products and ionic strengths were "adjusted to infinite dilution by iteration, to allow for incomplete dissociation..."

AUXILIARY INFORMATION

METHOD/APPARATUS/PROCEDURE:

Potentionmetric titration of 0.01 mol dm^{-3} $NaBPh_4$ with 0.01 mol dm^{-3} $AgClO_4$. The ionic strength at the midpoint of the titration curve was used to calculate the activity coefficient.

SOURCE AND PURITY OF MATERIALS:

Not stated.

ESTIMATED ERROR:

Nothing specified. A precision of ±0.1 pK units is assumed by the compiler.

REFERENCES:

COMPONENTS:
(1) Silver tetraphenylborate (1-); $AgC_{24}H_{20}B$; [14637-35-5]
(2) Sodium nitrate; $NaNO_3$ [7631-99-4]
(3) Sodium tetraphenylborate; $NaC_{24}H_{20}B$; [143-66-8]
(4) Formamide; CH_3NO; [75-12-7]

ORIGINAL MEASUREMENTS:
Alexander, R.; Ko, E. C. F.; Mac, Y. C.; Parker, A. J. *J. Am. Chem. Soc.* 1967, *89*, 3703-12.

VARIABLES:
One temperature: 25°C

PREPARED BY:
Orest Popovych

EXPERIMENTAL VALUES:

The solubility product of $AgBPh_4$ in formamide was reported as a product of ionic concentrations determined at ionic strengths in the range of 0.01-0.05 mol dm^{-3}, maintained by $NaBPh_4$:

$$pK^{\circ}_{s0} = 10.3 \ (K^{\circ}_{s0} \text{ units are in mol}^2 \text{ dm}^{-6}).$$

AUXILIARY INFORMATION

METHOD/APPARATUS/PROCEDURE:
Potentiometric titration of 0.01 mol dm^{-3} solution of $NaBPh_4$ with $AgNO_3$ using a silver indicator electrode and a silver-silver nitrate reference cell connected by a bridge of NEt_4 picrate. A radiometer pH meter, Type PHM22r was used. Cell was thermostatted, but limits of temperature control were not specified.

SOURCE AND PURITY OF MATERIALS:
Formamide was dried with Type 4A molecular sieves and fractionated twice under a reduced pressure of dry nitrogen. The salts were of Analar grade and were used as received.

COMMENTS AND/OR ADDITIONAL DATA:
The above solubility product seems to comprise a combination of concentration (for the BPh_4^- ion) and activity (for the Ag^+ ion). The validity of the reported value is therefore doubtful.

ESTIMATED ERROR:
Nothing specified. A precision of ±0.1 pK units is assumed by the compiler.

REFERENCES:

COMPONENTS:

(1) Silver tetraphenylborate (1-); $AgC_{24}H_{20}B$; [14637-35-5]

(2) Hexamethylphosphorotriamide; $C_6H_{18}N_3OP$; [680-31-9]

ORIGINAL MEASUREMENTS:

Alexander, R.; Ko, E. C. F.: Mac, Y. C.; Parker, A. J. *J. Am. Chem. Soc.* 1967, *89*, 3703-12.

VARIABLES:

One temperature: 25°C

PREPARED BY:

Orest Popovych

EXPERIMENTAL VALUES:

The solubility product of $AgBPh_4$ in hexamethylphosphorotriamide was reported as a product of ionic concentrations at the ionic strength corresponding to the solubility:

$$pK_{s0} = 4.7 \ (K_{s0} \text{ units are mol}^2 \text{ dm}^{-6}).$$

AUXILIARY INFORMATION

METHOD/APPARATUS/PROCEDURE:

Not specified unambiguously: it could have been either potentiometric titration with KI or UV-spectrophotometry on a Unicam SP500 spectrophotometer. Saturated solutions were prepared by shaking for 24 hours at 35°C and then for a further 24 hours at 25°C.

SOURCE AND PURITY OF MATERIALS:

Hexamethylphosphorotriamide was dried with Type 4A molecular sieves and fractionated twice under a reduced pressure of dry nitrogen.

ESTIMATED ERROR:

Nothing specified. A precision of ±0.1 pK units is assumed by the compiler.

REFERENCES:

COMPONENTS:

(1) Silver tetraphenylborate (1-); $AgC_{24}H_{20}B$; [14627-35-5]

(2) Methanol; CH_4O; [67-56-1]

EVALUATOR:

Orest Popovych, Department of Chemistry, The City University of New York, Brooklyn College, Brooklyn, N. Y. 11210, U. S. A.
September 1979

CRITICAL EVALUATION:

Only two publications reported the solubility product of silver tetraphenylborate ($AgBPh_4$) in methanol, both at 298 K (1, 2). The formal (concentration) solubility product determined by Alexander et al. (1) from a potentiometric titration of BPh_4^- with $AgNO_3$ using a silver electrode at ionic strength kept constant at 0.01 mol dm^{-3} was expressed as pK_{s0} = 13.2 (all K_{s0} units in this evaluation are mol^2 dm^{-6}).

Kolthoff and Chantooni (2) employed both potentiometry in the presence of $NaBPh_4$ and chemical-exchange reactions in which either solid AgBr was equilibrated with a solution of $NaBPh_4$ or solid $AgBPh_4$ was equilibrated with a solution of NaBr to determine the pK_{s0}° of $AgBPh_4$. They obtained the same result by both methods: pK_{s0}° = 14.4 ± 0.01.

Despite the fact that the above result was obtained by two independent methods, one hesitates to designate it as better than tentative, because of the relatively poor precision of the determination. The corresponding solubility at 298 K, calculated simply as $(K_{s0}^{\circ})^{1/2}$ would be (6.3 ± 0.7) x 10^{-8} mol dm^{-3} (evaluator). From the results of the individual potentiometric experiments (see compilation), it is possible to calculate the K_{s0}° = (4.0 ± 0.8) x 10^{-15} and the corresponding solubility of (6.3 ± 0.6) x 10^{-8} mol dm^{-3} (compiler). The latter values of K_{s0}° and of the solubility are preferable to the extent that no error due to conversion from the pK value was involved. Nevertheless, it would seem that a relative precision of better than 10% could perhaps be achieved for the solubility in the future.

REFERENCES:

1. Alexander, R.; Ko, E. C. F.; Mac, Y. C.; Parker, A. J. *J. Am. Chem. Soc.* 1967, *89*, 3703.
2. Kolthoff, I. M.; Chantooni, M. K., Jr. *Anal. Chem.* 1972, *44*, 194.

COMPONENTS:

(1) Silver tetraphenylborate (1-); $AgC_{24}H_{20}B$; [14637-35-5]
(2) Sodium nitrate; $NaNO_3$; [7631-99-4]
(3) Sodium tetraphenylborate; $NaC_{24}H_{20}B$; [143-66-8]
(4) Methanol; CH_4O; [67-56-1]

ORIGINAL MEASUREMENTS:

Alexander, R.; Ko, E. C. F.; Mac, Y. C.; Parker, A. J. *J. Am. Chem. Soc.* 1967, *89*, 3703-12.

VARIABLES:

One temperature: 25°C

PREPARED BY:

Orest Popovych

EXPERIMENTAL VALUES:

The solubility product of $AgBPh_4$ in methanol was reported as a product of ionic concentrations determined at ionic strengths in the range of 0.01-0.005 mol dm^{-3} maintained by $NaBPh_4$:

$$pK_{s0} = 13.2 \text{ } (K_{s0} \text{ units are mol}^2 \text{ dm}^{-6}).$$

AUXILIARY INFORMATION

METHOD/APPARATUS/PROCEDURE:

Potentiometric titration of 0.01 mol dm^{-3} solution of $NaBPh_4$ with $AgNO_3$ using a silver indicator electrode and a silver-silver nitrate reference cell connected by a bridge of NEt_4 picrate. A radiometer pH meter, Type PHM22r was used. Cell was thermostatted, but limits of temperature control were not specified.

SOURCE AND PURITY OF MATERIALS:

For the purification method for methanol the reader was referred to a literature source (1).
The salts were of Analar grade and were used as received.

ESTIMATED ERROR:

None specified. A precision of ±0.1 pK units is assumed by the compiler.

REFERENCES:

(1) Clare, B. W.; Cook, D.; Ko, E. C. F., Mac, Y. C.; Parker, A. J. *J. Am. Chem. Soc.* 1966, *88*, 1911.

COMPONENTS:

(1) Silver tetraphenylborate (1-); $AgC_{24}H_{20}B$; [14627-35-5]
(2) Sodium tetraphenylborate; $NaC_{24}H_{20}B$; [143-66-8]
(3) Sodium bromide; NaBr; [7647-15-6]
(4) Methanol; CH_4O; [67-56-1]

ORIGINAL MEASUREMENTS:

Kolthoff, I. M.; Chantooni, M. K., Jr. *Anal. Chem.* 1972, *44*, 194-5.

VARIABLES:

One temperature: 25°C

PREPARED BY:

Orest Popovych

EXPERIMENTAL VALUES:

The solubility (ion-activity) product of $AgBPh_4$ in methanol was reported to be:

$$pK^\circ_{s0} = 14.4 \text{ } (K^\circ_{s0} \text{ units are mol}^2 \text{ dm}^{-6}).$$

The above value was derived from chemical-exchange experiments described by the equilibrium:

$$NaBPh_4 + AgBr(s) \rightleftarrows AgBPh_4(s) + NaBr$$

from which the ratio: $\dfrac{K^\circ_{s0}(AgBPh_4)}{K^\circ_{s0}(AgBr)} = \dfrac{[BPh_4^-]}{[Br^-]}$ was measured.

In order to calculate $pK^\circ_{s0}(AgBPh_4)$, the value of $pK^\circ_{s0}(AgBr) = 15.5$ was taken from the literature (1). For NaBr, partial ion pairing was corrected for using a literature value for the dissociation constant, $K^d = 0.10$ (2) (presumably in units of mol dm^{-3}). Activity coefficients y were calculated from the partially extended Debye-Hückel equation, using the following values for the ion-size parameters:

BPh_4^- -1.2 nm, Br^- -0.3 nm, Na^+ -0.4 nm.

The solubility product of $AgBPh_4$ was determined also from potentiometric measurements in $NaBPh_4$ solutions saturated with $AgBPh_4$, employing

(continued.....)

AUXILIARY INFORMATION

METHOD/APPARATUS/PROCEDURE: A suspension of AgBr in a solution of $NaBPh_4$ or a suspension of $AgBPh_4$ in a solution of NaBr were shaken under deaerated conditions for 5 days. The filtered saturated solutions were analyzed spectrophotometrically for the BPh_4^- ion in the 260-280-nm range and potentiometrically for the Br^- ion, using a silver electrode. Prior to the potentionmetry, the suspensions were filtered, the filtrate evaporated and the residue heated in a muffle furnace at 600°C for 2 hrs in order to destroy the BPh_4^-. The residue was taken up in 10 cm^3 of 0.1 mol dm^{-3} HNO_3, 40 cm^3 of methanol were added and the solution titrated with $AgNO_3$. $K_{s0}(AgBPh_4)$ was also determined potentiometrically from the potentials of an Ag electrode in $NaBPh_4$ solutions saturated with $AgBPh_4$. A 0.01 mol dm^{-3} $AgNO_3$/Ag electrode in methanol was the reference (no salt bridge specified). A Cary Model 15 spectrophotometer and a Corning Model 10 pH meter were the instruments employed.

SOURCE AND PURITY OF MATERIALS:

Methanol was Matheson Spectroquality distilled once over Mg turnings (H_2O content 0.01% by Karl Fischer). NaBr was Mallinckrodt AR Grade, recrystallized from water. $NaBPh_4$ was Aldrich puriss. product, recrystallized by a literature method (3). $AgBPh_4$ was prepared by metathesis of $AgNO_3$ and 2% excess of $NaBPh_4$. The product was washed well with methanol, dried *in vacuo* at 50°C for 3 hrs., all preparations and storage being in the dark. AgBr was prepared and handled analogously.

ESTIMATED ERROR:

Precision ±0.1 pK units
Accuracy ±0.2 pK units
Temperature control: ±1°C

REFERENCES:
(1) Buckley, P.; Hartley, H. *Phil. Mag.* 1929, *8*, 320.
(2) Jervis, R.; Muir, D.; Butler, J.; Gordon, A. *J. Am. Chem. Soc.* 1953, *75*, 2855.
(3) Popov, A. I.; Humphrey, R. *J. Am. Chem. Soc.* 1959, *81*, 2043.

COMPONENTS:	ORIGINAL MEASUREMENTS:
(1) Silver tetraphenylborate (1-); $AgC_{24}H_{20}B$; [14627-35-5] (2) Sodium tetraphenylborate; $NaC_{24}H_{20}B$; [143-66-8] (3) Sodium bromide; NaBr; [7647-15-6] (4) Methanol; CH_4O; [67-56-1]	Kolthoff, I. M.; Chantooni, M. K., Jr. *Anal. Chem.* 1972, *44*, 194-5.

EXPERIMENTAL VALUES: (continued.....)

a silver indicator electrode and a 0.01 mol dm^{-3} $AgNO_3/Ag$ reference electrode in methanol. It was reported as pK°_{s0} = 14.4 ± 0.1 (K°_{s0} units are mol^2 dm^{-6}). Below are the results of individual experiments.

Exchange Experiments

(All concentrations are in mol dm^{-3}. The quantities in brackets are equilibrium concentrations).

Initial C_{NaBPh_4} or C_{NaBr}	10^4[NaBr]	$10^4[Br^-]$	$10^2[BPh_4^-]$	$y_{BPh_4^-}$	y_{Br^-}	$K^\circ_{s0}(AgBPh_4)$ / $K^\circ_{s0}(AgBr)$	pK°_{s0} $(AgBPh_4)$
			$NaBPh_4$ + AgBr				
0.0030	0.37	7.0	0.86_0	0.78	0.72	12	14.4
0.0186	1.0	1.3_7	1.7_1	0.74	0.62	15	14.3
			NaBr + AgBr				
0.0113	0.77	$12._7$	1.00	0.77	0.70	8.7	14.5
0.0226	2.25	$27._2$	1.99	0.72	0.60	8.8	14.5

Potentiometric Experiments

(All concentrations and activities are in units of mol dm^{-3})

10^3 C_{NaBPh_4}	E, mV	10^{12} a_{Ag^+}	$y_{BPh_4^-}$	$K^\circ_{s0}(AgBPh_4) \times 10^{15}/mol^2\ dm^{-6}$
1.57	-540	2.5	0.88	3.5
3.92	-552	1.6	0.83	5.2
9.80	-586	0.43	0.78	3.3

The authors report only an average value of pK°_{s0}. This compiler calculates from the above data K°_{s0} = (4.0 ± 0.8) x 10^{-15} mol^2 dm^{-6}. Neglecting activity corrections, the solubility is (6.3 ± 0.6) x 10^{-8} mol dm^{-3} (compiler).

COMPONENTS:

(1) Silver tetraphenylborate (1-); $AgC_{24}H_{20}B$; [14637-35-5]
(2) Sodium perchlorate; $NaClO_4$; [7601-89-0]
(3) Sodium tetraphenylborate; $NaC_{24}H_{20}B$; [143-66-8]
(4) N-Methyl-2-pyrrolidinone; C_5H_9NO; [872-50-4]

ORIGINAL MEASUREMENTS:

Alexander, R.; Parker, A. J.; Sharp, J. H.; Waghorne, W. E. *J. Am. Chem. Soc.* 1972, *94*, 1148-58.

VARIABLES:

One temperature: 25°C

PREPARED BY:

Orest Popovych

EXPERIMENTAL VALUES:

The solubility (ion-activity) product of $AgBPh_4$ in N-methyl-2-pyrrolidinone was determined in the presence of 0.01-0.005 mol dm^{-3} $NaBPh_4$ and reported as:

$$pK^{\circ}_{s0} = 4.9 \ (K^{\circ}_{s0} \text{ units are mol}^2 \text{ dm}^{-6}).$$

The mean ionic acticity coefficient was calculated from the Davies equation in the form: $\log \gamma_{\pm} = -A\,[(I)^{1/2}/(1 + (I)^{1/2}) - (1/3)I]$, where $\underline{I}$ is the ionic strength in mol dm^{-3} and the value of $\underline{A}$ used was 2.004 $mol^{-1/2}$ $dm^{3/2}$.

AUXILIARY INFORMATION

METHOD/APPARATUS/PROCEDURE:

Potentionmetric titration of 0.01 mol dm^{-3} $NaBPh_4$ with 0.01 mol dm^{-3} $AgClO_4$. The ionic strength at the midpoint of the titration curve was used to calculate the activity coefficient.

SOURCE AND PURITY OF MATERIALS:

Not stated.

ESTIMATED ERROR:

Nothing specified. A precision of ±0.1 pK units is assumed by the compiler.

REFERENCES:

COMPONENTS:

(1) Silver tetraphenylborate (1-); $AgC_{24}H_{20}B$; [14637-35-5]
(2) Sodium perchlorate; $NaClO_4$; [7601-89-0]
(3) Sodium tetraphenylborate; $NaC_{24}H_{20}B$; [143-66-8]
(4) Nitromethane; CH_3NO_2; [75-52-5]

ORIGINAL MEASUREMENTS:

Alexander, R.; Parker, A. J.; Sharp, J. H.; Waghorne, W. E. *J. Am. Chem. Soc.* 1972, *94*, 1148-58.

VARIABLES:

One temperature: 25°C

PREPARED BY:

Orest Popovych

EXPERIMENTAL VALUES:

The solubility (ion-activity) product of $AgBPh_4$ in nitromethane was determined in the presence of 0.01-0.005 mol dm^{-3} $NaBPh_4$ and reported as:

$$pK^{\circ}_{s0} = 15.6 \quad (K^{\circ}_{s0} \text{ units are } mol^2\ dm^{-6}).$$

The mean ionic activity coefficient was calculated from the Davies equation in the form: $\log \gamma_{\pm} = -A\ [(I)^{1/2}/(1 + (I)^{1/2}) - (1/3)\ I]$, where $\underline{I}$ is the ionic strength in mol dm^{-3} and the value of $\underline{A}$ used was 1.479 $mol^{-1/2}\ dm^{3/2}$.

AUXILIARY INFORMATION

METHOD/APPARATUS/PROCEDURE:

Potentiometric tetration of 0.01 mol dm^{-3} $NaBPh_4$ with 0.01 mol dm^{-3} $AgClO_4$. The ionic strength at the midpoint of the titration curve was used to calculate the activity coefficient.

SOURCE AND PURITY OF MATERIALS:

Not stated.

ESTIMATED ERROR:

Nothing specified. A precision of ±0.1 pK units is assumed by the compiler.

REFERENCES:

COMPONENTS:
(1) Silver tetraphenylborate (1-); $AgC_{24}H_{20}B$; [14637-35-5]
(2) Sodium perchlorate; $NaClO_4$; [7601-89-0]
(3) Sodium tetraphenylborate; $NaC_{24}H_{20}B$; [143-66-8]
(4) Propanediol-1,2-carbonate (propylene carbonate); $C_4H_6O_3$; [108-32-7]

ORIGINAL MEASUREMENTS:
Alexander, R.; Parker, A. J.; Sharp, J. H.; Waghorne, W. E. *J. Am. Chem. Soc.* 1972, *94*, 1148-58.

VARIABLES:
One temperature: 25°C

PREPARED BY:
Orest Popovych

EXPERIMENTAL VALUES:

The solubility (ion-activity) product of $AgBPh_4$ in propylene carbonate was determined in the presence of 0.01-0.005 mol dm^{-3} $NaBPh_4$ and reported as:

$$pK^{\circ}_{s0} = 12.8 \ (K^{\circ}_{s0} \text{ units are mol}^2 \text{ dm}^{-6}).$$

The mean ionic activity coefficient was calculated from the Davies equation in the form: $\log \gamma_{\pm} = -A\ [(I)^{1/2}/(1 + (I)^{1/2}) - (1/3)I]$, where I is the ionic strength in mol dm^{-3} and the value of A used was 0.661 $mol^{-1/2}\ dm^{3/2}$.

AUXILIARY INFORMATION

METHOD/APPARATUS/PROCEDURE:

Potentiometric titration of 0.01 mol dm^{-3} $NaBPh_4$ with 0.01 mol dm^{-3} $AgClO_4$. The ionic strength at the midpoint of the titration curve was used to calculate the activity coefficient.

SOURCE AND PURITY OF MATERIALS:

Not stated.

ESTIMATED ERROR:

Nothing specified. A precision of ±0.1 pK units is assumed by the compiler.

REFERENCES:

COMPONENTS:

(1) Silver tetraphenylborate (1-); $AgC_{24}H_{20}B$; [14637-35-5]
(2) Sodium perchlorate; $NaClO_4$; [7601-89-0]
(3) Sodium tetraphenylborate; $NaC_{24}H_{20}B$; [143-66-8]
(4) 2-Propanone (acetone); C_3H_6O; [67-64-1]

ORIGINAL MEASUREMENTS:

Alexander, R.; Parker, A. J.; Sharp, J. H. Waghorne, W. E. *J. Am. Chem. Soc.* 1972, *94*, 1148-58.

VARIABLES:

One temperature: 25°C

PREPARED BY:

Orest Popovych

EXPERIMENTAL VALUES:

The solubility (ion-activity) product of $AgBPh_4$ in acetone was determined in the presence of 0.01-0.005 mol dm^{-3} $NaBPh_4$ and reported as:

$$pK^{\circ}_{s0} = 13.1 \quad (K^{\circ}_{s0} \text{ units are mol}^2 \text{ dm}^{-6}).$$

The mean ionic activity coefficient was calculated from the Davies equation in the form: $\log \gamma_{\pm} = -A\,[(I)^{1/2}/(1 + (I)^{1/2}) - (1/3)I]$, where $\underline{I}$ is the ionic strength in mol dm^{-3} and the value of $\underline{A}$ used was 3.760 $mol^{-1/2}$ $dm^{3/2}$. The solubility products and ionic strengths were "adjusted to infinite dilution by iteration, to allow for incomplete dissociation..."

AUXILIARY INFORMATION

METHOD/APPARATUS/PROCEDURE:

Potentiometric titration of 0.01 mol dm^{-3} $NaBPh_4$ with 0.01 mol dm^{-3} $AgClO_4$. The ionic strength at the midpoint of the titration curve was used to calculate the activity coefficient.

SOURCE AND PURITY OF MATERIALS:

Not stated.

ESTIMATED ERROR:

Nothing specified. A precision of ±0.1 pK units is assumed by the compiler.

REFERENCES:

COMPONENTS:

(1) Silver tetraphenylborate (1-); $AgC_{24}H_{20}B$; [14637-35-5]

(2) Tetrahydrothiophene-1,1-dioxide (sulfolane, tetramethylene sulfone); $C_4H_8O_2S$; [126-33-0]

EVALUATOR:

Orest Popovych, Department of Chemistry, The City University of New York, Brooklyn College, Brooklyn, N. Y. 11210, U. S. A.
September 1979

CRITICAL EVALUATION:

The solubility product of silver tetraphenylborate ($AgBPh_4$) in sulfolane at 303 K was reported twice from the same laboratory (1, 2). The first determination employed UV-spectrophotometry, obtaining a formal (concentration) solubility product expressed as $pK_{s0} = 9.5 \pm 0.1$ (here all K_{s0} values have units of $mol^2\ dm^{-6}$). In the second publication (2) the determination was based on a potentiometric titration of BPh_4^- with Ag^+ using a silver electrode, coupled with a calculation of activity coefficients from the Davies equation (see compilation). The new result was $pK^\circ_{s0} = 10.2$. Unfortunately, neither study mentioned the nature of temperature control and the second study did not specify how saturation was ascertained. The solubility itself might be best calculated from the concentration pK_{s0} of 9.5, because it is at least free of uncertainty with respect to the activity correction. Taking simply the square root of the K_{s0}, we obtain for the solubility: $(1.8 \pm 0.2) \times 10^{-5}\ mol\ dm^{-3}$ as the tentative value.

REFERENCES:

1. Parker, A. J.; Alexander, R. *J. Am. Chem. Soc.* 1968, *90*, 3313.
2. Alexander, R.; Parker, A. J.; Sharp, J. H.; Waghorne, W. E. *J. Am. Chem. Soc.* 1972, *94*, 1148.

COMPONENTS:

(1) Silver tetraphenylborate(1-); $AgC_{24}H_{20}B$; [14637-35-5]

(2) Tetrahydrothiophene-1,1-dioxide (sulfolane); $C_4H_8O_2S$; [126-33-0]

ORIGINAL MEASUREMENTS:

Parker, A. J.; Alexander, R.
J. Am. Chem. Soc. 1968, *90*, 3313-9.

VARIABLES:

One temperature: 30°C

PREPARED BY:

Orest Popovych

EXPERIMENTAL VALUES:

The formal (concentration) solubility product of $AgBPh_4$ in sulfolane was reported as:

$$pK^\circ_{s0} = 9.5 \ (K^\circ_{s0} \text{ units are mol}^2 \text{ dm}^{-6}).$$

AUXILIARY INFORMATION

METHOD/APPARATUS/PROCEDURE:

UV spectrophotometry on solutions saturated under nitrogen, using a Unicam SP500 spectrophotometer. Saturated solutions were prepared by shaking for 24 hours at 35°C and then further for 24 hours at 30°C.

SOURCE AND PURITY OF MATERIALS:

The purification of materials has been described in the literature (1-3).

ESTIMATED ERROR:

Absolute precision was estimated to be ±0.1 pK units.

REFERENCES:

(1) Clare, B. W.; Cook, D.; Ko, E. C. F.; Mac, Y. C.; Parker, A. J. *J. Am. Chem. Soc.* 1966, *88*, 1911.
(2) Alexander, R.; Ko, E. C. F.; Mac, Y. C.; Parker, A. J. *J. Am. Chem. Soc.* 1967, *89*, 3703.
(3) Parker, A. J. *J. Chem. Soc. A* 1966, 220.

COMPONENTS:
(1) Silver tetraphenylborate (1-); $AgC_{24}H_{20}B$; [14637-35-5]
(2) Sodium perchlorate; $NaClO_4$; [7601-89-0]
(3) Sodium tetraphenylborate; $NaC_{24}H_{20}B$; [143-66-8]
(4) Tetrahydrothiophene-1,1-dioxide (sulfolane); $C_4H_8O_2S$; [126-33-0]

ORIGINAL MEASUREMENTS:
Alexander, R.; Parker, A. J.; Sharp, J. H.; Waghorne, W. E. *J. Am. Chem. Soc.* 1972, *94*, 1148-58.

VARIABLES:
One temperature: 30°C

PREPARED BY:
Orest Popovych

EXPERIMENTAL VALUES:

The solubility (ion-activity) product of $AgBPh_4$ in sulfolane was determined in the presence of 0.01-0.005 mol dm^{-3} $NaBPh_4$ and reported as:

$$pK^{\circ}_{s0} = 10.2 \quad (K^{\circ}_{s0} \text{ units are mol}^2 \text{ dm}^{-6}).$$

The mean ionic activity coefficient was calculated from the Davies equation in the form: $\log \gamma_{\pm} = -A\,[(I)^{1/2}/(1 + (I)^{1/2}) - (1/3)I]$, where I is the ionic strength in mol dm^{-3} and the value of A used was 1.244 $mol^{-1/2}$ $dm^{3/2}$.

AUXILIARY INFORMATION

METHOD/APPARATUS/PROCEDURE:

Potentiometric titration of 0.01 mol dm^{-3} $NaBPh_4$ with 0.01 mol dm^{-3} $AgClO_4$. The ionic strength at the midpoint of the titration curve was used to calculate the activity coefficient.

SOURCE AND PURITY OF MATERIALS:

Not stated.

ESTIMATED ERROR:

Not specified.
A precision of ±0.1 pK units is assumed by the compiler.

REFERENCES:

COMPONENTS:

(1) Thallium(I) tetraphenylborate (1-); $TlC_{24}H_{20}B$; [14637-31-1]

(2) Water; H_2O; [7732-18-5]

EVALUATOR:

Orest Popovych, Department of Chemistry, City University of New York, Brooklyn College, Brooklyn, N. Y. 11210, U. S. A.
December 1979

CRITICAL EVALUATION:

The solubility of thallium(I) tetraphenylborate ($TlBPh_4$) at 298 K was reported as 5.29×10^{-5} mol dm^{-3} in pure water (1) and as 1.1×10^{-5} mol dm^{-3} in a THAM buffer solution (2). Both were determined by uv-spectrophotometry, but the molar absorption coefficients in the first study were characteristic of acetonitrile (not aqueous) solutions. (For a discussion of the consequences, see the critical evaluation for $KBPh_4$ in aqueous systems). In the second study, the molar absorption coefficients were not specified.

Because the ionic strength of the THAM buffer in the second study is not known, it is impossible to estimate the corresponding solubility at zero ionic strength. However, it is clear that the latter value should be lower than 1.1×10^{-5} mol dm^{-3}, which makes for an even greater discrepancy between the two literature data. Thus, the solubility of 5.29×10^{-5} mol dm^{-3} reported by Pflaum and Howick (1) must be evaluated as highly tentative.

REFERNECES:

1. Pflaum, R. T.; Howick, L. C. *Anal. Chem.* 1956, *28*, 1542.
2. McClure, J. E.; Rechnitz, G. A. *Anal. Chem.* 1966, *38*, 136.

COMPONENTS:

(1) Thallium(I) tetraphenylborate (1-); $TlC_{24}H_{20}B$; [14637-31-1]

(2) Water; H_2O; [7732-18-5]

ORIGINAL MEASUREMENTS:

Pflaum, R. T.; Howick, L. C. *Anal. Chem.* 1956, *28*, 1542-4.

VARIABLES:

One temperature: 25°C

PREPARED BY:

Orest Popovych

EXPERIMENTAL VALUES:

The solubility of $TlBPh_4$ in water was reported as 5.29×10^{-5} mol dm^{-3}.

AUXILIARY INFORMATION

METHOD/APPARATUS/PROCEDURE:

Ultraviolet spectrophotometry. For details see the compilation for $KBPh_4$ in water based on the same reference.

SOURCE AND PURITY OF MATERIALS:

See compilation for $KBPh_4$ in water based on the same reference. TlBPh was prepared by metathesis of TlCl and $NaBPh_4$ and recrystallized from an acetonitrile-water mixture.

ESTIMATED ERROR:

Nothing is specified, but the precision is likely to be ±1% (compiler).

REFERENCES:

COMPONENTS:

(1) Thallium(I) tetraphenylborate (1-); $TlC_{24}H_{20}B$; [14637-31-1]
(2) Tris(hydroxymethyl)aminoethane; $C_4H_{11}NO_3$; [77-86-1]
(3) Acetic acid; $C_2H_4O_2$; [64-19-7]
(4) Water; H_2O; [7732-18-5]

ORIGINAL MEASUREMENTS:

McClure, J. E.; Rechnitz, G. A.
Anal. Chem. 1966, *38*, 136-9.

VARIABLES:

One temperature: 24.8°C

PREPARED BY:

Orest Popovych

EXPERIMENTAL VALUES:

The solubility of thallium(I) tetraphenylborate ($TlBPh_4$) in aqueous tris(hydroxymethyl)aminomethane (THAM) at pH 5.1 was reported as:

$$1.1 \times 10^{-5}\ \text{mol dm}^{-3}.$$

AUXILIARY INFORMATION

METHOD/APPARATUS/PROCEDURE:

UV-spectrophotometry according to the procedure of Howick and Pflaum (1). No other details.

SOURCE AND PURITY OF MATERIALS:

The buffer solution consisted of 0.1 mol dm^{-3} THAM and 0.01 mol dm^{-3} acetic acid, adjusted to pH 5.1 with G. F. Smith reagent-grade $HClO_4$. The source of BPh_4^- was a solution of $Ca(BPh_4)$ in THAM prepared from Fisher Scientific reagent-grade $NaBPh_4$ by the procedure of Rechnitz et al. (2) and standardized by potentiometric titration with KCl and RbCl. Tl^+ solutions were prepared by dissolving Tl_2CO_3 (A. D. Mackay, Inc.) in $HClO_4$.

ESTIMATED ERROR:

Not stated.
Temperature: ±0.3°C

REFERENCES:

(1) Howick, L. C.; Pflaum, R. T. *Anal. Chim. Acta* 1958, *19*, 342.
(2) Rechnitz, G. A.; Katz, S. A.; Zamochnick, S. B. *Anal. Chem.* 1963, *35*, 1322.

COMPONENTS:

(1) Thallium(I) tetraphenylborate (1-); $TlC_{24}H_{20}B$; [14637-31-1]

(2) N,N-Dimethylformamide; C_3H_7NO; [68-12-2]

ORIGINAL MEASUREMENTS:

Kolthoff, I. M.; Chantooni, M. K., Jr. *J. Phys. Chem.* 1972, *76*, 2024-34.

VARIABLES:

One temperature: 25°C

PREPARED BY:

Orest Popovych

EXPERIMENTAL VALUES:

The solubility product of $TlBPh_4$ in N,N-dimethylformamide was reported as:

$$pK^\circ_{s0} = 4.5 \ (K^\circ_{s0} \text{ units are mol}^2 \text{ dm}^{-6}).$$

The above value was derived from the electrolytic conductivity of the saturated solution $\kappa = 3.80 \times 10^{-4}$ ohm^{-1} cm^{-1} and a mean molar activity coefficient calculated from the Guggenheim equation (not shown).

AUXILIARY INFORMATION

METHOD/APPARATUS/PROCEDURE:

Electrolytic conductance of the saturated solution, using apparatus previously described (1). The authors state the precision of the conductance data as ±2%. The value for the λ^∞ of Tl^+ employed in the calculation was taken from the literature (2) and if the average value from the above source was used, it was 91.1 S cm^2 mol^{-1}. The λ^∞ for the BPh_4^- ion was estimated from the Walden rule using the known (unspecified) value in acetonitrile. Presumably, the solubility $\underline{C}$ was calculated using the relationship $C = 1000\kappa/\Lambda^\infty$, but this is not explained in the text and the actual solubility is not reported.

COMMENTS: Walden's rule is known to be unreliable. The use of limiting conductivities for solutions of the order of 10^{-2} mol dm^{-3} is questionable. The method of ascertaining saturation was not specified.

SOURCE AND PURITY OF MATERIALS:

N,N-Dimethylformamide was purified by a literature method (3). $TlBPh_4$ was prepared as described in the compilation for dimethysulfoxide.

ESTIMATED ERROR:

Nothing specified.

REFERENCES:

(1) Kolthoff, I. M.; Bruckenstein, S.; Chantooni, M. K., Jr. *J. Am. Chem. Soc.* 1961, *83*, 3927.

(2) Yeager, H. L.; Kratochvil, B. *J. Phys. Chem.* 1970, *74*, 963.

(3) Kolthoff, I. M.; Chantooni, M. K., Jr.; Smagowski, H. *Anal. Chem.* 1970, *42*, 1622.

COMPONENTS:

(1) Thallium(I) tetraphenylborate (1-); $TlC_{24}H_{20}B$; [14637-31-1]

(2) Dimethylsulfoxide; C_2H_6OS; [67-68-5]

ORIGINAL MEASUREMENTS:

Kolthoff, I. M.; Chantooni, M. K., Jr. *J. Phys. Chem.* 1972, *76*, 2024-34.

VARIABLES:

One temperature: 25°C

PREPARED BY:

Orest Popovych

EXPERIMENTAL VALUES:

The solubility of $TlBPh_4$ in dimethylsulfoxide was reported as:

$$C = 4.9 \times 10^{-2} \text{ mol dm}^{-3}.$$

After determining that the above salt was completely dissociated (based on conductance data), the authors calculated the mean ionic activity coefficients from the Guggenheim equation (not shown) and reported the solubility product of $TlBPh_4$ in dimethylsulfoxide as:

$$pK^\circ_{s0} = 2.9 \ (K^\circ_{s0} \text{ units are mol}^2 \text{ dm}^{-6}).$$

AUXILIARY INFORMATION

METHOD/APPARATUS/PROCEDURE:

Electrolytic conductance of the saturated solution, using previously described apparatus (1). The electrolytic conductivity of the saturated solution at 25°C was reported as 11.5×10^{-4} ohm^{-1} cm^{-1}, with a precision of ±2%. Values of Λ^∞ in dimethylsulfoxide were estimated from those in acetonitrile using Walden's rule. Presumably, the solubility C was calculated from the relationship $C = 1000\kappa/\Lambda^\infty$, where κ is the electrolytic conductivity.

SOURCE AND PURITY OF MATERIALS:

Dimethylsulfoxide was thoroughly purified by a literature method (2). Sodium tetraphenylborate (Aldrich puriss. grade) was purified by the method of Popov and Humphrey (3). $TlBPh_4$ was prepared by metathesis of $TlNO_3$ with $NaBPh_4$.

ESTIMATED ERROR:

Nothing specified.

COMMENTS:

Walden's rule is notoriously unreliable. In addition, errors are incurred by employing limiting conductivities at concentrations as high as the reported solubility value. The method of ascertaining saturation was not specified.

REFERENCES:

(1) Kolthoff, I. M.; Bruckenstein, S.; Chantooni, M. K., Jr. *J. Am. Chem. Soc.* 1961, *83*, 3927.
(2) Kolthoff, I. M.; Reddy, T. B. *Inorg. Chem.* 1962, *1*, 189.
(3) Popov, A. I.; Humphrey, R. *J. Am. Chem. Soc.* 1959, *81*, 2043.

COMPONENTS:	EVALUATOR:
(1) Tetraphenylarsonium tetraphenylborate (1-); $C_{48}H_{40}BAs$; [15627-12-0] (2) Water; H_2O; [7732-18-5]	Orest Popovych, Department of Chemistry, City University of New York, Brooklyn College, Brooklyn, N. Y. 11210, U. S. A. November 1979

CRITICAL EVALUATION:

All three reasonable estimates of the solubility of tetraphenylarsonium tetraphenylborate ($Ph_4As\ BPh_4$) in water come from Parker's laboratory (1-3) and all were determined at 298 K. The first two were reported as formal (concentration) solubility products (in the units of $mol^2\ dm^{-6}$) expressed in the form pK_{s0} = 16.7 (1) and 17.2 (2), respectively. The latter value corresponds to a solubility of 2.5 x 10^{-9} mol dm^{-3}, but the authors' estimate of precision is ±0.5 pK units, which means that the solubility could range anywhere from 4.5 x 10^{-9} mol dm^{-3} to 1.4 x 10^{-9} mol dm^{-3}. Although the spectral region in which the saturated solutions were analyzed was not specified, the uv-determination of the solubility must have been carried out in the non-specific short-wavelength region of the near-uv spectrum, where absorption is high, but where the tetraphenyl species cannot be distinguished from other aromatics, including decomposition products. On the other hand, in the region of 260-275 nm, where tetraphenyl compounds show characteristic spectra, the molar absorption coefficients are of the order of 10^3 $dm^3 (cm\ mol)^{-1}$, which renders impossible a uv-analysis of 10^{-9} mol dm^{-3} solutions. The need to analyze in the non-specific region of the spectrum may be responsible for the low precision of the reported solubility product.

Subsequently, Cox and Parker (3) expressed their preference for the determination of the solubility product by chemical-exchange experiments, between a solution of $AgNO_3$ and solid $Ph_4As\ BPh_4$ and a solution of $Ph_4As\ NO_3$ and solid $AgBPh_4$. The resulting value of pK_{s0} = 17.4 (K_{s0} units are $mol^2\ dm^{-6}$) is difficult to assess as to precision. For one thing, it is based on the literature value for the pK_{s0} of $AgBPh_4$ which itself is subject to an error of the order of 0.1 pK units (4). Furthermore, Cox and Parker did not mention whether or not any corrections were introduced for the activity coefficients of Ag^+ and Ph_4As^+ ions in their computation. Therefore, the K_{s0} of $Ph_4As\ BPh_4$ is likely to be partially based on activity and partially on concentration. Since the precision in the pK_{s0} can be no better than ±0.2 pK units, the solubility of $Ph_4As\ BPh_4$ in water calculated as $(K_{s0})^{1/2}$ could range from about 2.5 x 10^{-9} mol dm^{-3} to 1.6 x 10^{-9} mol dm^{-3}. The nominal solubility value of 2.0 x 10^{-9} mol dm^{-3} should be regarded therefore as no better than tentative.

In view of the results in all three studies by Parker et al. (1-3), the K_{s0} value of 5.0 x 10^{-9} $mol^2\ dm^{-6}$ reported by Cole and Pflaum (5) must be of the wrong order of magnitude. Since the authors did specify uv-analysis at 264 nm and 271 nm, it would appear that what they observed was absorption due to decomposition products and possibly, the starting materials.

References:

1. Alexander, R.; Parker, A. J. *J. Am. Chem. Soc.* 1967, *89*, 5549.
2. Parker, A. J.; Alexander, R. *J. Am. Chem. Soc.* 1968, *90*, 3313.
3. Cox, B. G.; Parker, A. J. *J. Am. Chem. Soc.* 1972, *94*, 3674.
4. Kolthoff, I. M.; Chantooni, M. K., Jr. *Anal. Chem.* 1972, *44*, 194.
5. Cole, J. J.; Pflaum, R. T. *Proc. Iowa Acad. Sciences* 1964, *71*, 145.

COMPONENTS:	ORIGINAL MEASUREMENTS:
(1) Tetraphenylarsonium tetraphenylborate (1-); $C_{48}H_{40}BAs$; [15627-12-0] (2) Water; H_2O; [7732-18-5]	Alexander, R.; Parker, A. J. *J. Am. Chem. Soc.* 1967, *89*, 5549-51.
VARIABLES:	PREPARED BY:
One temperature: 25°C	Orest Popovych

EXPERIMENTAL VALUES:

The formal (concentration) solubility product of $Ph_4As\ BPh_4$ in water was reported as:

$$pK_{s0} = 16.7 \ (K_{s0} \text{ units are mol}^2 \text{ dm}^{-6}).$$

AUXILIARY INFORMATION

METHOD/APPARATUS/PROCEDURE:	SOURCE AND PURITY OF MATERIALS:
UV spectrophotometry on solutions saturated under nitrogen. No other details.	Not stated.
	ESTIMATED ERROR:
	None specified.
	REFERENCES:

COMPONENTS:	ORIGINAL MEASUREMENTS:
(1) Tetraphenylarsonium tetraphenylborate (1-); $C_{48}H_{40}BAs$; [15627-12-0] (2) Water; H_2O; [7732-18-5]	Parker, A. J.; Alexander, R. *J. Am. Chem. Soc.* 1968, *90*, 3313-9.
VARIABLES:	PREPARED BY:
One temperature: 25°C	Orest Popovych

EXPERIMENTAL VALUES:

The formal (concentration) solubility product of $Ph_4As\ BPh_4$ in water was reported as:

$$pK_{s0} = 17.2 \ (K_{s0} \text{ units are mol}^2 \text{ dm}^{-6}).$$

AUXILIARY INFORMATION

METHOD/APPARATUS/PROCEDURE:

UV spectrophotometry on solutions saturated under nitrogen, using a Unicam SP500 spectrophotometer. Saturated solutions were prepared by shaking for 24 hours at 35°C and then for a further 24 hours at 25°C.

SOURCE AND PURITY OF MATERIALS:

The purification of materials has been described in the literature (1-3).

ESTIMATED ERROR:

Absolute precision was estimated to be ±0.5 pK units.

REFERENCES: (1) Clare, B. W.; Cook, D.; Ko, E. C. F.; Mac, Y. C.; Parker, A. J. *J. Am. Chem. Soc.* 1966, *88*, 1911.
(2) Alexander, R.; Ko, E. C. F.; Mac, Y. C.; Parker, A. J. *J. Am. Chem. Soc.* 1967, *89*, 3703.
(3) Parker, A. J. *J. Am. Chem. Soc. A* 1966, 220.

COMPONENTS:
(1) Tetraphenylarsonium tetraphenylborate (1-); $C_{48}H_{40}BAs$; [15627-12-0]
(2) Silver nitrate; $AgNO_3$; [7761-88-8]
(3) Tetraphenylarsonium nitrate; $C_{24}H_{20}AsNO_3$; [6727-90-8]
(4) Water; H_2O; [7732-18-5]

ORIGINAL MEASUREMENTS:
Cox, B. G.; Parker, A. J.
J. Am. Chem. Soc. 1972, *94*, 3674-5.

VARIABLES:
One temperature: 25°C

PREPARED BY:
Orest Popovych

EXPERIMENTAL VALUES:

The solubility product of $Ph_4As\ BPh_4$ in water was reported in the form:

$$pK_{s0} = 17.4 \ (K_{s0} \text{ units are mol}^2 \text{ dm}^{-6}).$$

The above value was obtained from measurements on the equilibrium:

$$Ph_4As\ BPh_4\ (s) + AgNO_3 \rightleftarrows Ph_4As\ NO_3 + AgBPh_4\ (s)$$

which is governed by the relationship:

$$pK_{s0}(Ph_4As\ BPh_4) = -\log\ [(Ph_4As^+)/(Ag^+)] + pK_{s0}(AgBPh_4)$$

The value of $pK_{s0}(AgBPh_4) = 17.2$ was taken from the literature (1).

AUXILIARY INFORMATION

METHOD/APPARATUS/PROCEDURE:
Fifty cm^3 of 0.01 mol dm^{-3} $AgNO_3$ were equilibrated with 1 g of $Ph_4As\ BPh_4$ containing a trace of $AgBPh_4$ as seed and 50 cm^3 of 0.01 mol dm^{-3} $Ph_4As\ NO_3$ were equilibrated with 1 g of $AgBPh_4$ containing a trace of $Ph_4As\ BPh_4$ as seed, all in CO_2-free water under nitrogen and in light-proof vessels. The solutions were analyzed for Ag^+ by atomic absorption and for Ph_4As^+ by UV spectrophotometry at 265 nm.

SOURCE AND PURITY OF MATERIALS:
Not stated.

ESTIMATED ERROR:
Not specified.

REFERENCES:
(1) Kolthoff, I. M.; Chantooni, M. K. Jr. *Anal. Chem.* 1972, *44*, 194.

COMPONENTS:

(1) Tetraphenylarsonium tetraphenylborate (1-); $C_{48}H_{40}BAs$; [15627-12-0]

(2) Water; H_2O; [7732-18-5]

ORIGINAL MEASUREMENTS:

Cole, J. J.; Pflaum, R. T. *Proc. Iowa Acad. Sciences* 1964, *71*, 145-150.

VARIABLES:

One temperature: 25.0°C

PREPARED BY:

Orest Popovych

EXPERIMENTAL VALUES:

The authors reported the solubility of $Ph_4As\ BPh_4$ in water at 25°C as:

$C = 4.99 \times 10^{-4}$ g/100 cm^3 and the corresponding K_{s0} as 5.0×10^{-9} $mol^2\ dm^{-6}$. Apparently K_{s0} was calculated as C^2, i.e., there was no activity correction.

AUXILIARY INFORMATION

METHOD/APPARATUS/PROCEDURE:

UV spectrophotometry at 264 and 271 nm using a Cary Model 14 spectrophotometer. Saturated solutions prepared in a constant-temperature bath and equilibration was assumed when successive analyzes agreed to within ±0.5%.

SOURCE AND PURITY OF MATERIALS:

Ph_4AsCl (G. Frederick Smith Chemical Co.) and $NaBPh_4$ were reacted to form the $Ph_4As\ BPh_4$, which was recrystallized from acetone-water.

ESTIMATED ERROR:

Precision of ±0.5% (authors).
Temperature control: ±0.1°C

REFERENCES:

COMPONENTS:	EVALUATOR:
(1) Tetraphenylarsonium tetraphenylborate (1-); $C_{48}H_{40}BAs$; [15627-12-0] (2) Acetonitrile; C_2H_3N; [75-05-8]	Orest Popovych, Department of Chemistry, City University of New York, Brooklyn College, Brooklyn, N. Y. 11210, U. S. A. November 1979

CRITICAL EVALUATION:

All four literature values pertaining to the solubility of tetraphenylarsonium tetraphenylborate ($Ph_4As\ BPh_4$) in acetonitrile were determined at 298 K by Parker and his associates (1-4). All data were reported as pK_{s0} values where the K_{s0} units were $mol^2\ dm^{-6}$. The first datum, pK_{s0} = 5.2 (1) was reported with a paucity of experimental detail, except that is was a concentration solubility product determined by uv-spectrophotometry. It was superceded by another concentration solubility product, expressed as pK_{s0} = 5.7 with the precision stated as ±0.1 pK units (2). Since the experimental details were somewhat better defined in this second article, the above value may serve as the basis for calculating the solubility as $(K_{s0})^{\frac{1}{2}} = (1.4 \pm 0.2) \times 10^{-3}\ mol\ dm^{-3}$. Considering that the solubility value is precise to only one decimal and that no mention is made in the article of ascertaining saturation, this datum must be considered as only tentative.

The two later data are less reliable as sources of the solubility value. The thermodynamic ion-activity product reported as pK°_{s0} = 5.8 (3) contained an activity coefficient calculated from the Davies equation (see compilation), but the break-down between the solubility and the activity coefficient was not shown. Finally, Cox and Parker (4) expressed preference for determining the solubility product of $Ph_4As\ BPh_4$ from chemical-exchange experiments, rather than by direct uv-spectrophotometry (see compilation). Unfortunately, it is not clear whether activity coefficients were used in the calculation of the solubility product, so that the reported pK_{s0} = 6.0 may be a concentration or an activity product and therefore not suitable for the calculation of the solubility value. The precision of the last pK_{s0} value cannot be better than ±0.2 pK units, considering that it is based on a literature value of the pK_{s0} ($AgBPh_4$) which itself is precise to only ±0.1 pK units.

References:

1. Alexander, R; Parker. A. J. *J. Am. Chem Soc.* 1967, *89*, 5549.
2. Parker, A. J.; Alexander, R. *J. Am. Chem Soc.* 1968, *90*, 3313.
3. Alexander, R.; Parker, A. J. Sharp, J. H.; Waghorne, W. E. *J. Am. Chem Soc.* 1972, *94*, 1148.
4. Cox, B. G.; Parker, A. J. *J. Am. Chem. Soc.* 1972, *94*, 3674.

COMPONENTS:

(1) Tetraphenylarsonium tetraphenylborate (1-); $C_{48}H_{40}BAs$; [15627-12-0]

(2) Acetonitrile; C_2H_3N; [75-05-8]

ORIGINAL MEASUREMENTS:

Alexander, R.; Parker, A. J.
J. Am. Chem. Soc. 1967, *89*, 5549-51.

VARIABLES:

One temperature: 25°C

PREPARED BY:

Orest Popovych

EXPERIMENTAL VALUES:

The formal (concentration) solubility product of $Ph_4As\ BPh_4$ in acetonitrile was reported as:

$$pK_{s0} = 5.2 \ (K_{s0} \text{ units are mol}^2 \text{ dm}^{-6}).$$

AUXILIARY INFORMATION

METHOD/APPARATUS/PROCEDURE:

UV spectrophotometry on solutions saturated under nitrogen. No other details.

SOURCE AND PURITY OF MATERIALS:

Not stated.

ESTIMATED ERROR:

None specified.

REFERENCES:

COMPONENTS:	ORIGINAL MEASUREMENTS:
(1) Tetraphenylarsonium tetraphenylborate (1-); $C_{48}H_{40}BAs$; [15627-12-0] (2) Acetonitrile; C_2H_3N; [75-05-8]	Parker, A. J.; Alexander, R. *J. Am. Chem. Soc.* 1968, *90*, 3313-9.
VARIABLES:	PREPARED BY:
One temperature: 25°C	Orest Popovych

EXPERIMENTAL VALUES:

The formal (concentration) solubility product of $Ph_4As\ BPh_4$ in acetonitrile was reported as:

$$pK_{s0} = 5.7 \ (K_{s0} \text{ units are mol}^2 \text{ dm}^{-6})$$

AUXILIARY INFORMATION

METHOD/APPARATUS/PROCEDURE:

UV spectrophotometry on solutions saturated under nitrogen, using a Unicam SP500 spectrophotometer. Saturated solutions were prepared by shaking for 24 hours at 35°C and then for a further 24 hours at 25°C.

SOURCE AND PURITY OF MATERIALS:

The purification of materials has been described in the literature (1-3).

ESTIMATED ERROR:

Absolute precision was estimated to be ±0.1 pK units.

REFERENCES:

(1) Clare, B. W.; Cook, D.; Ko, E. C. F.; Mac, Y. C.; Parker, A. J. *J. Am. Chem. Soc.* 1966, *88*, 1911.
(2) Alexander, R.; Ko, E. C. F.; Mac, Y. C.; Parker, A. J. *J. Am. Chem. Soc.* 1967, *89*, 3703.
(3) Parker, A. J. *J. Chem. Soc. A* 1966, 220.

COMPONENTS:

(1) Tetraphenylarsonium tetraphenylborate (1-); $C_{40}H_{48}As$; [15627-12-0]

(2) Acetonitrile; C_2H_3N; [75-05-8]

ORIGINAL MEASUREMENTS:

Alexander, R.; Parker, A. J.; Sharp, J. H.; Waghorne, W. E. *J. Am. Chem. Soc.* 1972, *94*, 1148-58.

VARIABLES:

One temperature: 25 °C

PREPARED BY:

Orest Popovych

EXPERIMENTAL VALUES:

The solubility (ion-activity) product of $Ph_4As\ BPh_4$ in acetonitrile was reported as:

$$pK^{\circ}_{s0} = 5.8 \ (K^{\circ}_{s0} \text{ units are mol}^2 \text{ dm}^{-6}).$$

The mean ionic activity coefficient was calculated from the Davies equation in the form: $\log \gamma_{\pm} = -A\ [(I)^{1/2}/(1 + (I)^{1/2}) - (1/3)I]$, where $\underline{I}$ is the ionic strength in mol dm^{-3} and the value of $\underline{A}$ used was 1.543 $mol^{-1/2}\ dm^{3/2}$.

AUXILIARY INFORMATION

METHOD/APPARATUS/PROCEDURE:

Probably UV spectrophotometry. No other details.

SOURCE AND PURITY OF MATERIALS:

Not stated.

ESTIMATED ERROR:

Not specified.

REFERENCES:

COMPONENTS:

(1) Tetraphenylarsonium tetraphenylborate (1-); $C_{48}H_{40}BAs$; [15627-12-0]
(2) Silver nitrate; $AgNO_3$; [7761-88-8]
(3) Tetraphenylarsonium nitrate; $C_{24}H_{20}AsNO_3$; [6727-90-8]
(4) Acetonitrile; C_2H_3N; [75-05-8]

ORIGINAL MEASUREMENTS:

Cox, B. G.; Parker, A, J.
J. Am. Chem. Soc. 1972, *94*, 3674-5.

VARIABLES:

One temperature: 25°C

PREPARED BY:

Orest Popovych

EXPERIMENTAL VALUES:

The solubility product of $Ph_4As\ BPh_4$ in acetonitrile was reported in the form;

$$pK_{s0} = 6.0 \ (K_{s0} \text{ units are mol}^2 \text{ dm}^{-6}).$$

The above value was derived from measurements on the equilibrium:

$$Ph_4As\ BPh_4\ (s) + AgNO_3 \rightleftarrows Ph_4As\ NO_3 + AgBPh_4\ (s)$$

which is governed by the relationship:

$$pK_{s0}(Ph_4As\ BPh_4) = -\log[(Ph_4As^+)/(Ag^+)] + pK_{s0}(AgBPh_4)$$

The required value of $pK_{s0}(AgBPh_4) = 7.6$ was taken from the literature (1).

AUXILIARY INFORMATION

METHOD/APPARATUS/PROCEDURE:

Fifty cm^3 of 0.01 mol dm^{-3} $AgNO_3$ were equilibrated with 1 g of $Ph_4As\ BPh_4$ containing a trace of $AgBPh_4$ used as a seed and 50 cm^3 of 0.01 mol dm^{-3} $Ph_4As\ NO_3$ were equilibrated with 1 g of $AgBPh_4$ containing a trace of $Ph_4As\ BPh_4$ as seed, in anhydrous acetonitrile, under nitrogen in light-proof vessels. The solutions were analyzed for Ag^+ by atomic absorption and for Ph_4As^+ by UV spectrophotometry at 265 nm.

SOURCE AND PURITY OF MATERIALS:

Not stated.

ESTIMATED ERROR:

Not specified.

REFERENCES:

(1) Alexander, R.; Parker, A. J.; Sharp, J. H.; Waghorne, W. E. *J. Am. Chem. Soc.* 1972, *94*, 1143, and private communication from I. M. Kolthoff and M. K. Chantooni Jr.

COMPONENTS:

(1) Tetraphenylarsonium tetraphenylborate (1-); $C_{48}H_{40}BAs$; [15627-12-0]

(2) 1,1-Dichloroethane; $C_2H_4Cl_2$; [75-34-3]

ORIGINAL MEASUREMENTS:

Abraham, M. H.; Danil de Manor, A. F. *J. Chem. Soc. Faraday Trans. 1* 1976, *72*, 955-62.

VARIABLES:

One temperature: 25°C

PREPARED BY:

Orest Popovych

EXPERIMENTAL VALUES:

The authors reported the solubility of Ph_4AsBPh_4 in 1,1-dichloroethane as:
2.70×10^{-4} mol dm^{-3}.
Using an estimated association constant of 3.60×10^3 mol^{-1} dm^3 and an ion-size parameter of $\mathring{a} = 0.66$ nm with which to calculate the mean ionic activity coefficient from the extended Debye-Hückel equation, they obtained for the standard Gibbs free energy of solution:
$\Delta G^\circ_{s0} = 10.51$ kcal mol^{-1} = 43.99 kJ mol^{-1} (compiler).
The solubility (ion-activity) product of Ph_4AsBPh_4 can be calculated from the relationship $\Delta G^\circ_s = -RT \ln K^\circ_{s0}$, yielding $pK^\circ_{s0} = 7.705$, where K°_{s0} units are mol^2 dm^{-6} (compiler).

AUXILIARY INFORMATION

METHOD/APPARATUS/PROCEDURE:

Evaporation and weighing. Saturated solutions prepared by shaking the suspensions for several days at 25°C. No solvate was detected. Method of temperature control was not specified.

SOURCE AND PURITY OF MATERIALS:

The solvent was shaken with anhydrous K_2CO_3, passed through a column of basic activated alumina into distillation flask and fractionated under N_2 through a 3-foot column. At least 10% of distillate was rejected, the rest collected over freshly activated molecular sieve. The source and purification of Ph_4AsBPh_4 were not specified.

ESTIMATED ERROR:

Precision of 0.1 kcal mol^{-1} in ΔG°_s.

REFERENCES:

COMPONENTS:

(1) Tetraphenylarsonium tetraphenylborate (1−); $C_{48}H_{40}BAs$; [15627-12-0]

(2) 1,2-Dichloroethane; $C_2H_4Cl_2$; [107-06-2]

ORIGINAL MEASUREMENTS:

Abraham, M. H.; Danil de Namor, A. F. *J. Chem. Soc. Faraday Trans. 1* 1976, *72*, 955-62.

VARIABLES:

One temperature: 25°C

PREPARED BY:

Orest Popovych

EXPERIMENTAL VALUES:

The authors reported the solubility of Ph_4AsBPh_4 in 1,2-dichloroethane as:
4.99×10^{-3} mol dm^{-3}.
Using an estimated association constant of 6.00×10^2 mol^{-1} dm^3 and an ion-size parameter of $\mathring{a} = 0.66$ nm with which to calculate the mean ionic activity coefficient from the extended Debye-Hückel equation, they obtained the standard Gibbs free energy of solution:

$$\Delta G_s^\circ = 7.90 \text{ kcal mol}^{-1} = 33.1 \text{ kJ mol}^{-1} \text{ (compiler)}.$$

The solubility (ion-activity) product of Ph_4AsBPh_4 can be calculated from the relationship:
$\Delta G_s^\circ = -RT \ln K_{s0}^\circ$, yielding $pK_{s0}^\circ = 5.792$, where K_{s0}° units are mol^2 dm^{-6} (compiler).

AUXILIARY INFORMATION

METHOD/APPARATUS/PROCEDURE:

Evaporation and weighing. Saturated solutions prepared by shaking the suspensions for several days at 25°C. No solvate was detected. Method of temperature control was not specified.

SOURCE AND PURITY OF MATERIALS:

The solvent was shaken with anhydrous K_2CO_3, passed through a column of basic activated alumina into distillation flask and fractionated under N_2 through a 3-foot column. At least 10% of distillate was rejected, the rest collected over freshly activated molecular sieve. The source and purification of Ph_4AsBPh_4 were not specified.

ESTIMATED ERROR:

Precision of 0.1 kcal mol^{-1} in ΔG_s°.

REFERENCES:

COMPONENTS:	EVALUATOR:
(1) Tetraphenylarsonium tetraphenylborate (1-); $C_{48}H_{40}BAs$; [15627-12-0] (2) N,N-Dimethylacetamide; C_4H_9NO; [127-19-5]	Orest Popovych, Department of Chemistry, City University of New York, Brooklyn College, Brooklyn, N. Y. 11210, U. S. A. November 1979

CRITICAL EVALUATION:

All three available data on the solubility of tetraphenylarsonium tetraphenylborate ($Ph_4As\ BPh_4$) in N,N-dimethylacetamide come from the laboratory of Parker and his associates (1-3). All were determined by uv-spectrophotometry at 298 K. The first two data were concentration solubility products (in $mol^2\ dm^{-6}$) reported in the form pK_{s0} = 3.4 (1) and 3.7 (2), respectively. Since the latter was obtained under conditions that were better defined in the article, with a specified precision of ±0.1 pK units, it represents the best available quantity from which the solubility can be estimated. Thus, if the solubility is taken simply as $(K_{s0})^{1/2}$, we obtain for it $(1.4 \pm 0.2) \times 10^{-2}\ mol\ dm^{-3}$. It should be considered as a tentative value. The thermodynamic solubility product determined subsequently and expressed as pK°_{s0} = 4.0 (3) is based on an activity coefficient calculated from the Davies equation (see compilation), but unfortunately the solubility itself was not specified in that study.

References:

1. Alexander, R.; Parker, A. J. *J. Am. Chem. Soc.* 1967, *89*, 5549.
2. Parker, A. J.; Alexander, R. *J. Am. Chem. Soc.* 1968, *90*, 3313.
3. Alexander, R.; Parker, A. J.; Sharp, J. H.; Waghorne, W. E. *J. Am. Chem. Soc.* 1972, *94*, 1148.

COMPONENTS:

(1) Tetraphenylarsonium tetraphenylborate (1-); $C_{48}H_{40}BAs$; [15627-12-0]

(2) N,N-Dimethylacetamide; C_4H_9NO; [127-19-5]

ORIGINAL MEASUREMENTS:

Alexander, R.; Parker, A. J.

J. Am. Chem. Soc. 1967, *89*, 5549-51.

VARIABLES:

One temperature: 25°C

PREPARED BY:

Orest Popovych

EXPERIMENTAL VALUES:

The formal (concentration) solubility product of $Ph_4As\ BPh_4$ in N,N-dimethylacetamide was reported as:

$$pK_{s0} = 3.4\ (K_{s0}\ \text{units are mol}^2\ \text{dm}^{-6}).$$

AUXILIARY INFORMATION

METHOD/APPARATUS/PROCEDURE:

UV spectrophotometry on solutions saturated under nitrogen.
No other details.

SOURCE AND PURITY OF MATERIALS:

Not stated.

ESTIMATED ERROR:

None specified.

REFERENCES:

COMPONENTS:	ORIGINAL MEASUREMENTS:
(1) Tetraphenylarsonium tetraphenylborate (1-); $C_{48}H_{40}BAs$; [15627-12-0] (2) N,N-Dimethylacetamide; C_4H_9NO; [127-19-5]	Parker, A. J.; Alexander, R. *J. Am. Chem. Soc.* 1968, *90*, 3313-9.
VARIABLES:	PREPARED BY:
One temperature: 25°C	Orest Popovych

EXPERIMENTAL VALUES:

The formal (concentration) solubility product of $Ph_4As\ BPh_4$ in N,N-dimethylacetamide was reported as:

$$pK_{s0} = 3.7 \ (K_{s0} \text{ units are mol}^2 \text{ dm}^{-6}).$$

AUXILIARY INFORMATION

METHOD/APPARATUS/PROCEDURE:

UV spectrophotometry on solutions saturated under nitrogen, using a Unicam SP500 spectrophotometer. Saturated solutions were prepared by shaking for 24 hours at 35°C and then for a further 24 hours at 25°C.

SOURCE AND PURITY OF MATERIALS:

The purification of materials has been described in the literature (1-3).

ESTIMATED ERROR:

Absolute precision was estimated to be ±0.1 pK units.

REFERENCES: (1) Clare, B. W.; Cook, D.; Ko, E. C. F.; Mac, Y. C.; Parker, A. J. *J. Am. Chem. Soc.* 1966, *88*, 1911.
(2) Alexander, R.; Ko, E. C. F.; Mac, Y. C.; Parker, A. J. *J. Am. Chem. Soc.* 1967, *89*, 3703.
(3) Parker, A. J. *J. Chem. Soc. A.* 1966, 220.

COMPONENTS:

(1) Tetraphenylarsonium tetraphenylborate (1-); $C_{48}H_{40}BAs$; [15627-12-0]

(2) N,N-Dimethylacetamide; C_4H_9NO; [127-19-5]

ORIGINAL MEASUREMENTS:

Alexander, R.; Parker, A. J.; Sharp, J. H.; Waghorne, W. E. *J. Am. Chem. Soc.* 1972, *94*, 1148-58.

VARIABLES:

One temperature: 25°C

PREPARED BY:

Orest Popovych

EXPERIMENTAL VALUES:

The solubility (ion-activity) product of $Ph_4As\ BPh_4$ in N,N-dimethylacetamide was reported as:

$$pK_{s0}^{\circ} = 4.0 \ (K_{s0}^{\circ} \text{ units are mol}^2 \text{ dm}^{-6}).$$

The mean ionic activity coefficient was calculated from the Davies equation in the form; $\log \gamma_{\pm} = -A\,[(I)^{\frac{1}{2}}/(1 + (I)^{\frac{1}{2}}) - (1/3)I]$, where $\underline{I}$ is the ionic strength in mol dm^{-3} and the value of $\underline{A}$ used was 1.551 $mol^{-1/2}\ dm^{3/2}$.

AUXILIARY INFORMATION

METHOD/APPARATUS/PROCEDURE:

Probably UV spectrophotometry.
No other details.

SOURCE AND PURITY OF MATERIALS:

Not stated.

ESTIMATED ERROR:

Not specified.

REFERENCES:

COMPONENTS:

(1) Tetraphenylarsonium tetraphenylborate (1-); $C_{48}H_{40}BAs$; [15627-12-0]

(2) N,N-Dimethylformamide; C_3H_7NO; [68-12-2]

EVALUATOR:

Orest Popovych, Department of Chemistry, City University of New York, Brooklyn College, Brooklyn, N. Y. 11210, U. S. A.
November, 1979

CRITICAL EVALUATION:

Three of the four available literature data pertaining to the solubility of tetraphenylarsonium tetraphenylborate ($Ph_4As\ BPh_4$) in N,N-dimethylformamide were determined at 298 K by uv-spectrophotometry in the laboratory of Parker and his associates (1-3). The first two data were concentration solubility products (in $mol^2\ dm^{-6}$) reported in the form pK_{s0} = 3.7 (1,2), showing agreement to one decimal in the logarithm between the two studies. If we take the solubility as being equal to $(K_{s0})^{\frac{1}{2}}$ and accept the authors' estimate of the precision as ±0.1 pK units (2), the solubility value becomes $(1.4 \pm 0.2) \times 10^{-2}$ mol dm^{-3}. The above value is probably the best estimate of the solubility we have to date, but it should be considered as no better than <u>tentative</u>.

The thermodynamic solubility product reported subsequently as pK°_{s0} = 4.9 (3) is not a reliable source of the solubility value because it contains an appreciable activity correction calculated via the Davies equation (see compilation). Unfortunately, the solubility was not specified separately in the last study (3).

The pK°_{s0} value of 3.9 reported by Kolthoff and Chantooni (4) on the basis of measurements of electrolytic conductance would seem less reliable for two reasons. Firstly, the values of molar conductances for the Ph_4As^+ and BPh_4^- ions required for the calculation (see compilations) were not known in N,N-dimethylformamide and had to be estimated from those in acetonitrile via Walden's rule. The latter can lead to very serious errors, however. Secondly, at the 10^{-2} mol dm^{-3} concentration levels, additional errors result from the use of <u>limiting</u> molar conductances. These objections pertain to the evaluation of the value of the solubility product. To derive the solubility from it, one would have to know the form of the Guggenheim equation used to calculate the activity coefficient, which was not specified by the authors.

<u>References</u>:

1. Alexander, R.; Parker, A. J. *J. Am. Chem. Soc.* <u>1967</u>, *89*, 5549.
2. Parker, A. J. ; Alexander, R. *J. Am. Chem. Soc.* <u>1968</u>, *90*, 3313.
3. Alexander, R.; Parker, A. J.; Sharp, J. H.; Waghorne, W. E. *J. Am. Chem. Soc.* <u>1972</u>, *94*, 1148.
4. Kolthoff, I. M.; Chantooni, M. K., Jr. *J. Phys. Chem.* <u>1972</u>, *76*, 2024.

COMPONENTS:	ORIGINAL MEASUREMENTS:
(1) Tetraphenylarsonium tetraphenylborate (1-); $C_{48}H_{40}BAs$; [15627-12-0] (2) N,N-Dimethylformamide; C_3H_7NO; [68-12-2]	Alexander, R.; Parker, A. J. *J. Am. Chem. Soc.* 1967, *89*, 5549-51.
VARIABLES:	PREPARED BY:
One temperature: 25°C	Orest Popovych

EXPERIMENTAL VALUES:

The formal (concentration) solubility product of $Ph_4As\ BPh_4$ in N,N-dimethylformamide was reported as:

$$pK_{s0} = 3.7\ (K_{s0} \text{ units are mol}^2\ dm^{-6}).$$

AUXILIARY INFORMATION

METHOD/APPARATUS/PROCEDURE:	SOURCE AND PURITY OF MATERIALS:
UV spectrophotometry on solutions saturated under nitrogen. No other details.	Not stated.
	ESTIMATED ERROR:
	None specified.
	REFERENCES:

COMPONENTS:

(1) Tetraphenylarsonium tetraphenylborate (1-); $C_{48}H_{40}BAs$; [15627-12-0]

(2) N,N-Dimethylformamide; C_3H_7NO; [68-12-2]

ORIGINAL MEASUREMENTS:

Parker, A. J.; Alexander, R.

J. Am. Chem. Soc. 1968, *90*, 3313-9.

VARIABLES:

One temperature: 25°C

PREPARED BY:

Orest Popovych

EXPERIMENTAL VALUES:

The formal (concentration) solubility product of $Ph_4As\ BPh_4$ in N,N-dimethylformamide was reported as:

$$pK_{s0} = 3.7 \quad (K_{s0} \text{ units are } mol^2\ dm^{-6}).$$

AUXILIARY INFORMATION

METHOD/APPARATUS/PROCEDURE:

UV spectrophotometry on solutions saturated under nitrogen, using a Unicam SP500 spectrophotometer. Saturated solutions were prepared by shaking for 24 hours at 35°C and then for a further 24 hours at 25°C.

SOURCE AND PURITY OF MATERIALS:

The purification of materials has been described in the literature (1-3).

ESTIMATED ERROR:

Absolute precision was estimated to be ± 0.1 pK units.

REFERENCES:

(1) Clare, B. W.; Cook, D.; Ko, E. C. F.; Mac, Y. C.; Parker, A. J. *J. Am. Chem. Soc.* 1966, *88*, 1911.
(2) Alexander, R.; Ko, E. C. F.; Mac, Y. C.; Parker. A. J. *J. Am. Chem. Soc.* 1967, *89*, 3703.
(3) Parker, A. J. *J. Chem. Soc. A* 1966, 220.

COMPONENTS:

(1) Tetraphenylarsonium tetraphenylborate (1-); $C_{48}H_{40}BAs$; [15627-12-0]

(2) N,N-Dimethylformamide; C_3H_7NO; [68-12-2]

ORIGINAL MEASUREMENTS:

Alexander, R.; Parker, A. J.; Sharp, J. H.; Waghorne, W. E. *J. Am. Chem. Soc.* 1972, *94*, 1148-58.

VARIABLES:

One temperature: 25°C

PREPARED BY:

Orest Popovych

EXPERIMENTAL VALUES:

The solubility (ion-activity) product of $Ph_4As\ BPh_4$ in N,N-dimethylformamide was reported as:

$$pK^{\circ}_{s0} = 4.9 \quad (K^{\circ}_{s0} \text{ units are mol}^2 \text{ dm}^{-6}).$$

The mean ionic activity coefficient was calculated from the Davies equation in the form: $\log \gamma_{\pm} = -A\ [(I)^{1/2}/(1 + (I)^{1/2}) - (1/3)I]$, where $\underline{I}$ is the ionic strength in mol dm^{-3} and the value of $\underline{A}$ used was 1.551 $mol^{-1/2}\ dm^{3/2}$.

AUXILIARY INFORMATION

METHOD/APPARATUS/PROCEDURE:

Probably UV spectrophotometry. No other details.

SOURCE AND PURITY OF MATERIALS:

Not stated.

ESTIMATED ERROR:

Not specified.

REFERENCES:

COMPONENTS:

(1) Tetraphenylarsonium tetraphenylborate (1-); $C_{48}H_{40}BAs$; [15627-12-0]

(2) N,N-Dimethylformamide; C_3H_7NO; [68-12-2]

ORIGINAL MEASUREMENTS:

Kolthoff, I. S.; Chantooni, M. K., Jr. *J. Phys. Chem.* 1972, *76*, 2024-34.

VARIABLES:

One temperature: 25°C

PREPARED BY:

Orest Popovych

EXPERIMENTAL VALUES:

The authors report only the solubility product of tetraphenylarsonium tetraphenylborate ($Ph_4As\ BPh_4$):

$$pK^{\circ}_{s0} = 3.9\ (K^{\circ}_{s0} \text{ has units of mol}^2\ \text{dm}^{-6}).$$

The above value was derived from the electrolytic conductivity of the saturated solution $\kappa = 5.00 \times 10^{-4}$ ohm^{-1} cm^{-1} and a mean molar activity coefficient calculated from the Guggenheim equation (not shown).

AUXILIARY INFORMATION

METHOD/APPARATUS/PROCEDURE:

Electrolytic conductance of the saturated solution, using conductance apparatus previously described (1). The authors state the precision of their conductance data as ±2%. The values of λ^{∞} for the BPh_4^- and the Ph_4As^+ ion required for the calculation of the solubility were estimated from those in acetonitrile and the Walden rule. Presumably the solubility C was obtained using the relationship $C = 1000\ \kappa/\Lambda^{\infty}$ but this is not explained and the actual C is not reported.

SOURCE AND PURITY OF MATERIALS:

N,N-Dimethylformamide was purified by a literature method (2). $Ph_4As\ BPh_4$ was prepared by the method of Popov and Humphrey (3).

ESTIMATED ERROR:

None specified.

REFERENCES:

(1) Kolthoff, I. M.; Bruckenstein, S. Chantooni, M. K., Jr. *J. Am. Chem. Soc.* 1961, *83*, 3927.

(2) Kolthoff, I. M.; Chantooni, M. K., Jr.; Smagowski, H. *Anal. Chem.* 1970, *42*, 1622.

(3) Popov, A. I.; Humphrey, R., *J. Am. Chem. Soc.* 1959, *81*, 2043.

COMPONENTS:

(1) Tetraphenylarsonium tetraphenylborate (1-); $C_{48}H_{40}BAs$; [15627-12-0]

(2) Dimethylsulfoxide C_2H_6OS; [67-68-5]

EVALUATOR:

Orest Popovych, Department of Chemistry, City University of New York, Brooklyn College, Brooklyn, N. Y. 11210, U. S. A.
November 1979

CRITICAL EVALUATION:

All four literature data pertaining to the solubility of tetraphenylarsonium tetraphenylborate ($Ph_4As\ BPh_4$) in dimethylsulfoxide are solubility products in units of $mol^2\ dm^{-6}$, reported in the form of pK_{s0} (or pK°_{s0}). All were determined at 298 K. The first two data were concentration solubility products determined by uv-spectrophotometry by Alexander and Parker and reported as pK_{s0} = 3.3 (1) and 3.6 (2), respectively. The second value was obtained under experimental conditions that were better described in the article and was accompanoed by a stated precision of ±0.1 pK units. Consequently, it may represent the best source for the solubility value, which calculated as $(K_{s0})^{1/2}$ is $(1.6 \pm 0.2) \times 10^{-2}$ mol dm^{-3} (evaluator) and should be regarded as the tentative value.

The thermodynamic solubility product reported subsequently as pK°_{s0} = 3.6 (3) was based on an activity coefficient calculated from the Davies equation (see compilation), but the break-down between the solubility and the activity correction was not shown. Considering the relatively high ionic concentration involved, the thermodynamic solubility product is not a good datum for the calculation of the solubility.

The only independent check on the above data from another laboratory comes from Kolthoff and Chantooni (4). Unfortunately, their value expressed as pK°_{s0} = 4.3 is subject to too many sources of error to be reliable. Here the solubility was calculated from electrolytic conductivity κ, presumably using the relationship Solubility = $1000\kappa/\Lambda^\infty$, where Λ^∞ is the limiting molar conductivity of the electrolyte. However, the limiting molar conductivity in dimethylsulfoxide was not known and had to be estimated from the Walden's rule, which is a highly unsatisfactory procedure. Furthermore, limiting molar conductivities are not anywhere near applicable to electrolyte solutions at concentrations of the order of 10^{-2} mol dm^{-3}.

References:

1. Alexander, R.; Parker, A. J. *J. Am. Chem. Soc.* 1967, *89*, 5549.
2. Parker, A. J.; Alexander, R. *J. Am. Chem. Soc.* 1968, *90*, 3313.
3. Alexander, R.; Parker, A. J.; Sharp, J. H.; Waghorne, W. E. *J. Am. Chem. Soc.* 1972, *94*, 1148.
4. Kolthoff, I. M.; Chantooni, M. K., Jr. *J. Phys. Chem.* 1972, *76*, 2024.

COMPONENTS:

(1) Tetraphenylarsonium tetraphenylborate (1-); $C_{48}H_{40}BAs$; [15627-12-0]

(2) Dimethylsulfoxide; C_2H_6OS; [67-68-5]

ORIGINAL MEASUREMENTS:

Alexander, R.; Parker, A. J. *J. Am. Chem. Soc.* 1967, *89*, 5549-51.

VARIABLES:

One temperature: 25°C

PREPARED BY:

Orest Popovych

EXPERIMENTAL VALUES:

The formal (concentration) solubility product of $Ph_4As\ BPh_4$ in dimethylsulfoxide was reported as:

$$pK_{s0} = 3.3 \ (K_{s0} \text{ units are mol}^2 \text{ dm}^{-6}).$$

AUXILIARY INFORMATION

METHOD/APPARATUS/PROCEDURE:

UV spectrophotometry on solutions saturated under nitrogen. No other details.

SOURCE AND PURITY OF MATERIALS:

Not stated.

ESTIMATED ERROR:

None specified.

REFERENCES:

COMPONENTS:	ORIGINAL MEASUREMENTS:
(1) Tetraphenylarsonium tetraphenylborate (1-); $C_{48}H_{40}BAs$; [15627-12-0] (2) Dimethylsulfoxide; C_2H_6OS; [67-68-5]	Parker, A. J.; Alexander, R. *J. Am. Chem. Soc.* 1968, *90*, 3313-9.
VARIABLES:	**PREPARED BY:**
One temperature: 25°C	Orest Popovych

EXPERIMENTAL VALUES:

The formal (concentration) solubility product of $Ph_4As\ BPh_4$ in dimethylsulfoxide was reported as:

$$pK_{s0} = 3.6 \ (K_{s0} \text{ units are mol}^2 \text{ dm}^{-6}).$$

AUXILIARY INFORMATION

METHOD/APPARATUS/PROCEDURE:

UV spectrophotometry on solutions saturated under nitrogen, using a Unicam SP500 spectrophotometer. Saturated solutions were prepared by shaking for 24 hours at 35°C and then for a further 24 hours at 25°C.

SOURCE AND PURITY OF MATERIALS:

The purification of materials has been described in the literature (1-3).

ESTIMATED ERROR:

Absolute precision was estimated to be ±0.1 pK units.

REFERENCES:

(1) Clare, B. W.; Cook, D.; Ko, E. C. F.; Mac, Y. C.; Parker, A. J. *J. Am. Chem Soc.* 1966, *88*, 1911.
(2) Alexander, R.; Ko, E. C. F.; Mac, Y. C.; Parker, A. J. *J. Am. Chem. Soc.* 1967, *89*, 3703.
(3) Parker, A. J. *J. Chem. Soc. A* 1966, 220.

COMPONENTS:

(1) Tetraphenylarsonium tetraphenylborate (1-); $C_{48}H_{40}BAs$; [15627-12-0]

(2) Dimethylsulfoxide; C_2H_6OS; [67-68-5]

ORIGINAL MEASUREMENTS:

Alexander, R.; Parker, A. J.; Sharp, J. H.; Waghorne, W. E. *J. Am. Chem. Soc.* 1976, *94*, 1148-58.

VARIABLES:

One temperature: 25°C

PREPARED BY:

Orest Popovych

EXPERIMENTAL VALUES:

The solubility (ion-activity) product of $Ph_4As\ BPh_4$ in dimethylsulfoxide was reported as:

$$pK^\circ_{s0} = 3.6 \ (K^\circ_{s0} \text{ units are mol}^2 \text{ dm}^{-6}).$$

The mean ionic activity coefficient was calculated from the Davies equation in the form: $\log \gamma_\pm = -A[(I)^{1/2}/(1 + (I)^{1/2}) - (1/3)I]$, where I is the ionic strength in mol dm^{-3} and the value of A used was 1.115 $mol^{-1/2}\ dm^{3/2}$.

AUXILIARY INFORMATION

METHOD/APPARATUS/PROCEDURE:

Probably UV spectrophotometry. No other details.

SOURCE AND PURITY OF MATERIALS:

Not stated.

ESTIMATED ERROR:

Not specified.

REFERENCES:

COMPONENTS:

(1) Tetraphenylarsonium tetraphenylborate (1-); $C_{48}H_{40}BAs$; [156-27-12-0]

(2) Dimethylsulfoxide; C_2H_6OS; [67-68-5]

ORIGINAL MEASUREMENTS:

Kolthoff, I. M.; Chantooni, M. K.,Jr. *J. Phys. Chem.* 1972, *76*, 2024-34.

VARIABLES:

One temperature: 25°C

PREPARED BY:

Orest Popovych

EXPERIMENTAL VALUES:

The authors report only the solubility product of tetraphenylarsonium tetraphenylborate ($Ph_4As\ BPh_4$):

$$pK_{s0}^{\circ} = 4.3 \quad (K_{s0}^{\circ} \text{ has units of } mol^2\ dm^{-6}).$$

The above value was derived from the electrolytic conductivity of the saturated solution $\kappa = 1.43 \times 10^{-4}\ ohm^{-1}\ cm^{-1}$ and a mean molar activity coefficient calculated from the Guggenheim equation (not shown).

AUXILIARY INFORMATION

METHOD/APPARATUS/PROCEDURE:

Electrolytic conductance. Identical to that described on the compilation for $Ph_4As\ BPh_4$ in N,N-dimethylformamide based on the same reference as this compilation.

SOURCE AND PURITY OF MATERIALS:

Dimethylsulfoxide was purified by a literature method (1). $Ph_4As\ BPh_4$ was prepared by the method of Popov and Humphrey (2).

ESTIMATED ERROR:

None specified.

REFERENCES:

(1) Kolthoff, I. M.; Reddy, T. B. *Inorg. Chem.* 1962, *1*, 189.

(2) Popov, A. I.; Humphrey, R. *J. Am. Chem. Soc.* 1959, *81*, 2043.

COMPONENTS:

(1) Tetraphenylarsonium tetraphenylborate (1-); $C_{48}H_{40}BAs$; [15627-12-0]

(2) Ethanol; C_2H_6O; [64-17-5]

ORIGINAL MEASUREMENTS:

Alexander, R.; Parker, A. J. Sharp, J. H.; Waghorne, W. E. *J. Am. Chem. Soc.* 1972, *94*, 1148-58.

VARIABLES:

One temperature: 25°C

PREPARED BY:

Orest Popovych

EXPERIMENTAL VALUES:

The solubility (ion-activity) product of $Ph_4As\ BPh_4$ in ethanol was reported as:

$$pK^\circ_{s0} = 9.4 \ (K^\circ_{s0} \text{ units are mol}^2 \text{ dm}^{-6}).$$

The mean ionic activity coefficient was calculated from the Davies equation in the form: $\log \gamma_\pm = -A\ [(I)^{1/2}/(1 + (I)^{1/2}) - (1/3)I]$, where I is the ionic strength in mol dm^{-3} and the value of A used was 2.956 $mol^{-1/2}\ dm^{3/2}$. The solubility products and ionic strengths were "adjusted to infinite dilution by iteration, to allow for incomplete dissociation..."

AUXILIARY INFORMATION

METHOD/APPARATUS/PROCEDURE:

Probably UV spectrophotometry. No other details.

SOURCE AND PURITY OF MATERIALS:

Not stated.

ESTIMATED ERROR:

Not specified.

REFERENCES:

COMPONENTS:	EVALUATOR:
(1) Tetraphenylarsonium tetraphenylborate (1-); $C_{48}H_{40}BAs$; [15627-12-0] (2) Formamide; CH_3NO; [75-12-7]	Orest Popovych, Department of Chemistry, City University of New York, Brooklyn College, Brooklyn, N. Y. 11210, U. S. A. November 1979

CRITICAL EVALUATION:

All three available data on the solubility of tetraphenylarsonium tetraphenylborate ($Ph_4As\ BPh_4$) in formamide were determined at 298 K by uv-spectrophotometry in the laboratory of Parker and his associates (1-3). The first two data were concentration solubility products (in $mol^2\ dm^{-6}$) reported in the form pK_{s0} = 8.3 (1) and 8.9 (2), respectively. Since the latter was obtained under experimental conditions that were better defined in the article, with a specified precision of ±0.1 pK units, it is the best available datum from which the solubility can be estimated as $(K_{s0})^{1/2}$. We obtain for the solubility (3.6 ± 0.4) x 10^{-5} mol dm^{-3} as a <u>tentative</u> value. Subsequently, the thermodynamic solubility product of $Ph_4As\ BPh_4$ in formamide was reported in the form pK°_{s0} = 8.8 (3). However, the solubility itself was not reported in the last study.

References:

1. Alexander, R.; Parker, A. J. *J. Am. Chem. Soc.* 1967, *89*, 5549.
2. Parker, A. J.; Alexander, R. *J. Am. Chem. Soc.* 1968, *90*, 3313.
3. Alexander, R.; Parker, A. J.; Sharp, J. H.; Waghorne, W. E. *J. Am. Chem. Soc.* 1972, *94*, 1148.

COMPONENTS:

(1) Tetraphenylarsonium tetraphenylborate (1-); $C_{48}H_{40}BAs$; [15627-12-0]

(2) Formamide; CH_3NO; [75-12-7]

ORIGINAL MEASUREMENTS:

Alexander, R.; Parker, A. J.
J. Am. Chem. Soc. 1967, *89*, 5549-51.

VARIABLES:

One temperature: 25°C

PREPARED BY:

Orest Popovych

EXPERIMENTAL VALUES:

The formal (concentration) solubility product of $Ph_4As\ BPh_4$ in formamide was reported as:

$$pK_{s0} = 8.3\ (K_{s0}\ \text{units are mol}^2\ \text{dm}^{-6}).$$

AUXILIARY INFORMATION

METHOD/APPARATUS/PROCEDURE:

UV spectrophotometry on solutions saturated under nitrogen.
No other details.

SOURCE AND PURITY OF MATERIALS:

Not stated.

ESTIMATED ERROR:

None specified.

REFERENCES:

COMPONENTS:

(1) Tetraphenylarsonium tetraphenylborate (1-); $C_{48}H_{40}BAs$; [15627-12-0]

(2) Formamide; CH_3NO; [75-12-7]

ORIGINAL MEASUREMENTS:

Parker, A. J.; Alexander, R. *J. Am. Chem. Soc.* 1968, *90*, 3313-9.

VARIABLES:

One temperature: 25°C

PREPARED BY:

Orest Popovych

EXPERIMENTAL VALUES:

The formal (concentration) solubility product of $Ph_4As\ BPh_4$ in formamide was reported as:

$$pK_{s0} = 8.9 \text{ } (K_{s0} \text{ units are mol}^2 \text{ dm}^{-6}).$$

AUXILIARY INFORMATION

METHOD/APPARATUS/PROCEDURE:

UV spectrophotometry on solutions saturated under nitrogen, using a Unicam SP500 spectrophotometer. Saturated solutions were prepared by shaking for 24 hours at 35°C and then for a further 24 hours at 25°C.

SOURCE AND PURITY OF MATERIALS:

The purification of materials has been described in the literature (1-3).

ESTIMATED ERROR:

Absolute precision was estimated to be ±0.1 pK units.

REFERENCES:

(1) Clare, B. W.; Cook, D.; Ko, E. C. F.; Mac, Y. C.; Parker, A. J. *J. Am. Chem. Soc.* 1966, *88*, 1911.
(2) Alexander, R.; Ko, E. C. F.; Mac, Y. C.; Parker, A. J. *J. Am. Chem. Soc.* 1967, *89*, 3703.
(3) Parker, A. J. *J. Chem. Soc. A* 1966, 220.

COMPONENTS:

(1) Tetraphenylarsonium tetraphenylborate (1-); $C_{48}H_{40}BAs$; [15627-12-0]

(2) Formamide; CH_3NO; [75-12-7]

ORIGINAL MEASUREMENTS:

Alexander, R.; Parker, A. J.; Sharp, J. S.; Waghorne, W. E. *J. Am. Chem. Soc.* 1972, *94*, 1148-58.

VARIABLES:

One temperature: 25°C

PREPARED BY:

Orest Popovych

EXPERIMENTAL VALUES:

The solubility (ion-activity) product of $Ph_4As\ BPh_4$ in formamide was reported as:

$$pK^{\circ}_{s0} = 8.8 \ (K^{\circ}_{s0} \text{ units are } mol^2\ dm^{-6}).$$

The mean ionic activity coefficient was calculated from the Davies equation in the form: $\log \gamma_{\pm} = -A\ [(I)^{1/2}/(1 + (I)^{1/2}) - (1/3)I]$, where $\underline{I}$ is the ionic strength in mol dm^{-3} and the value of $\underline{A}$ used was 0.309 $mol^{-1/2}\ dm^{3/2}$.

AUXILIARY INFORMATION

METHOD/APPARATUS/PROCEDURE:

Probably UV spectrophotometry.
No other details.

SOURCE AND PURITY OF MATERIALS:

Not stated.

ESTIMATED ERROR:

Not specified.

REFERENCES:

COMPONENTS:

(1) Tetraphenylarsonium tetraphenylborate (1-); $C_{48}H_{40}BAs$; [15627-12-0]

(2) Hexamethylphosphorotriamide; $C_6H_{18}N_3OP$; [680-31-9]

EVALUATOR:

Orest Popovych, Department of Chemistry, City University of New York, Brooklyn College, Brooklyn, N. Y. 11210, U. S. A.

November 1979

CRITICAL EVALUATION:

All three available data on the solubility of tetraphenylarsonium tetraphenylborate ($Ph_4As\ BPh_4$) in hexamethylphosphorotriamide were determined at 298 K by uv-spectrophotometry in the laboratory of Parker and his associates (1-3). The first two data were concentration solubility products (in $mol^2\ dm^{-6}$) reported in the form pK_{s0} = 3.1 (1) and 3.7 (2), respectively. Since the latter was obtained under experimental conditions that were better defined in the article, with a specified precision of ±0.1 pK units, it represents the best available quantity from which the solubility can be estimated. If we calculate the solubility simply as $(K_{s0})^{1/2}$, we obtain (1.4 ± 0.2) x 10^{-2} mol dm^{-3}. It should be considered as a tentative value. The thermodynamic solubility product reported subsequently as pK°_{s0} = 3.7 (3), which was calculated using an activity coefficient derived from the Davies equation (see compilation), has, surprisingly, the same numerical value as the earlier formal solubility product. Unless this is an error, it would imply that the thermodynamic solubility product was based on a new (revised) solubility value. The latter, unfortunately, was not reported.

References:

1. Alexander, R; Parker, A. J. *J. Am. Chem. Soc.* 1967, *89*, 5549.
2. Parker, A. J.; Alexander, R. *J. Am. Chem. Soc.* 1968, *90*, 3313.
3. Alexander, R.; Parker, A. J.; Sharp, J. H.; Waghorne, W. E. *J. Am. Chem Soc.* 1972, *94*, 1148.

COMPONENTS:	ORIGINAL MEASUREMENTS:
(1) Tetraphenylarsonium tetraphenylborate (1-); $C_{48}H_{40}BAs$; [15627-12-0] (2) Hexamethylphosphorotriamide; $C_6H_{18}N_3OP$; [680-31-9]	Alexander, R.; Parker, A. J. *J. Am. Chem. Soc.* 1967, *89*, 5549-51.
VARIABLES:	**PREPARED BY:**
One temperature: 25°C	Orest Popovych

EXPERIMENTAL VALUES:

The formal (concentration) solubility product of $Ph_4As\ BPh_4$ in hexamethylphosphorotriamide was reported as:

$$pK_{s0} = 3.1 \ (K_{s0} \text{ units are mol}^2 \text{ dm}^{-6}).$$

AUXILIARY INFORMATION

METHOD/APPARATUS/PROCEDURE:	SOURCE AND PURITY OF MATERIALS:
UV spectrophotometry on solutions saturated under nitrogen. No other details.	Not stated.
	ESTIMATED ERROR:
	None specified.
	REFERENCES:

COMPONENTS:

(1) Tetraphenylarsonium tetraphenylborate (1-); $C_{48}H_{40}BAs$; [15627-12-0]

(2) Hexamethylphosphorotriamide; $C_6H_{18}N_3OP$; [680-31-9]

ORIGINAL MEASUREMENTS:

Parker, A. J.; Alexander, R.

J. Am. Chem. Soc. 1968, *90*, 3313-9.

VARIABLES:

One temperature: 25°C

PREPARED BY:

Orest Popovych

EXPERIMENTAL VALUES:

The formal (concentration) solubility product of $Ph_4As\ BPh_4$ in hexamethylphosphorotriamide was reported as:

$$pK_{s0} = 3.7 \ (K_{s0} \text{ units are mol}^2 \text{ dm}^{-6}).$$

AUXILIARY INFORMATION

METHOD/APPARATUS/PROCEDURE:

UV spectrophotometry on solutions saturated under nitrogen, using a Unicam SP500 spectrophotometer. Saturated solutions were prepared by shaking for 24 hours at 35°C and then for a further 24 hours at 25°C.

SOURCE AND PURITY OF MATERIALS:

The purification of materials has been described in the literature (1-3).

ESTIMATED ERROR:

Absolute precision was estimated to be ±0.1 pK units.

REFERENCES:

(1) Clare, B. W.; Cook, D.; Ko, E. C. F.; Mac, Y. C.; Parker, A. J. *J. Am. Chem. Soc.* 1966, *88*, 1911.
(2) Alexander, R.; Ko, E. C. F., Mac, Y. C.; Parker, A. J. *J. Am. Chem. Soc.* 1967, *89*, 3703.
(3) Parker, A. J. *J. Chem. Soc. A.* 1966, 220.

COMPONENTS:

(1) Tetraphenylarsonium tetraphenylborate (1-); $C_{48}H_{40}BAs$; [15627-12-0]

(2) Hexamethylphosphorotriamide; $C_6H_{18}N_3OP$; [680-31-9]

ORIGINAL MEASUREMENTS:

Alexander, R.; Parker, A. J.; Sharp J. H.; Waghorne, W. E. *J. Am. Chem. Soc.* 1972, *94*, 1148-58.

VARIABLES:

One temperature: 25°C

PREPARED BY:

Orest Popovych

EXPERIMENTAL VALUES:

The solubility (ion-activity) product of $Ph_4As\ BPh_4$ in hexamethylphosphorotriamide was reported as:

$$pK^\circ_{s0} = 3.7 \quad (K^\circ_{s0} \text{ units are mol}^2\ \text{dm}^{-6}).$$

The mean ionic activity coefficient was calculated from the Davies equation in the form: $\log \gamma_\pm = -A\,[(I)^{1/2}/(1 + (I)^{1/2}) - (1/3)I]$, where I is the ionic strength in mol dm^{-3} and the value of A used was 2.201 $mol^{-1/2}\ dm^{3/2}$.

AUXILIARY INFORMATION

METHOD/APPARATUS/PROCEDURE:

Probably UV spectrophotometry. No other details.

SOURCE AND PURITY OF MATERIALS:

Not stated.

ESTIMATED ERROR:

Not specified.

REFERENCES:

COMPONENTS:	EVALUATOR:
(1) Tetraphenylarsonium tetraphenylborate (1-); $C_{48}H_{40}BAs$; [15627-12-0] (2) Methanol; CH_4O; [67-56-1]	Orest Popovych, Department of Chemistry, City University of New York, Brooklyn College, Brooklyn, N. Y. 11210, U. S. A. November 1979

CRITICAL EVALUATION:

All three determinations of the solubility of tetraphenylarsonium tetraphenylborate ($Ph_4As\ BPh_4$) in methanol were reported from the laboratory of Parker and his associates (1-3). All results were obtained at 298 K by uv-spectrophotometry. The first two data were reported to be concentration solubility products, expressed as pK_{s0} = 8.5 (1) and 9.0 (2), respectively. However, the 9.0 value was obtained under somewhat better defined experimental conditions and had a specified precision of ±0.1 pK units. Taking the solubility as $(K_{s0})^{\frac{1}{2}} = (10^{-9})^{\frac{1}{2}}$, we obtain for it the value (3.2 ± 0.4) x 10^{-5} mol dm^{-3}. Considering that temperature control was not specified and the estimate is good to only two significant figures, the above value cannot be considered as better than tentative.

The third datum was a thermodynamic solubility product for which the activity coefficient was calculated from the Davies equation shown in the compilation; it was expressed as pK°_{s0} = 9.0 (3). Unfortunately, the value of the solubility itself was not specified in the last study.

References:

1. Alexander, R.; Parker, A. J. *J. Am. Chem. Soc.* 1967, *89*, 5549.
2. Parker, A. J.; Alexander, R. *J. Am. Chem. Soc.* 1968, *90*, 3313.
3. Alexander, R.; Parker, A. J.; Sharp, J. H.; Waghorne, W. E. *J. Am. Chem. Soc.* 1972, *94*, 1148.

COMPONENTS:

(1) Tetraphenylarsonium tetraphenylborate (1-); $C_{48}H_{40}BAs$; [15627-12-0]

(2) Methanol; CH_4O; [67-56-1]

ORIGINAL MEASUREMENTS:

Alexander, R.; Parker, A. J. *J. Am. Chem. Soc.* 1967, *89*, 5549-51.

VARIABLES:

One temperature: 25°C

PREPARED BY:

Orest Popovych

EXPERIMENTAL VALUES:

The formal (concentration) solubility product of $Ph_4As\ BPh_4$ in methanol was reported as:

$$pK_{s0} = 8.5 \ (K_{s0} \text{ units are mol}^2 \text{ dm}^{-6}).$$

AUXILIARY INFORMATION

METHOD/APPARATUS/PROCEDURE:

UV spectrophotometry on solutions saturated under nitrogen.
No other details.

SOURCE AND PURITY OF MATERIALS:

Not stated.

ESTIMATED ERROR:

None specified.

REFERENCES:

COMPONENTS:

(1) Tetraphenylarsonium tetraphenylborate (1-); $C_{48}H_{40}BAs$; [15627-12-0]

(2) Methanol; CH_4O; [67-56-1]

ORIGINAL MEASUREMENTS:

Parker, A. J.; Alexander, R. *J. Am. Chem. Soc.* 1968, *90*, 3313-9.

VARIABLES:

One temperature: 25°C

PREPARED BY:

Orest Popovych

EXPERIMENTAL VALUES:

The formal (concentration) solubility product of $Ph_4As\ BPh_4$ in methanol was reported as:

$$pK_{s0} = 9.0 \ (K_{s0} \text{ units are mol}^2 \text{ dm}^{-6}).$$

AUXILIARY INFORMATION

METHOD/APPARATUS/PROCEDURE:

UV spectrophotometry on solutions saturated under nitrogen, using a Unicam SP500 spectrophotometer. Saturated solutions were prepared by shaking for 24 hours at 35°C and then for a further 24 hours at 25°C.

SOURCE AND PURITY OF MATERIALS:

The purification of materials has been described in the literature (1-3).

ESTIMATED ERROR:

Absolute precision was estimated to be ±0.1 pK units.

REFERENCES:

(1) Clare, B. W.; Cook, D.; Ko, E. C. F.; Mac, Y. C.; Parker, A. J. *J. Am. Chem. Soc.* 1966, *88*, 1911.
(2) Alexander, R.; Ko, E. C. F.; Mac, Y. C.; Parker, A. J. *J. Am. Chem. Soc.* 1967, *89*, 3703.
(3) Parker, A. J. *J. Chem. Soc. A* 1966, 220.

COMPONENTS:

(1) Tetraphenylarsonium tetraphenylborate (1-); $C_{48}H_{40}BAs$; [15627-12-0]

(2) Methanol; CH_4O; [67-56-1]

ORIGINAL MEASUREMENTS:

Alexander, R.; Parker, A. J.; Sharp, J. H.; Waghorne, W. E. *J. Am. Chem. Soc.* 1972, *94*, 1148-58.

VARIABLES:

One temperature: 25°C

PREPARED BY:

Orest Popovych

EXPERIMENTAL VALUES:

The solubility (ion-activity) product of $Ph_4As\ BPh_4$ in methanol was reported as:

$$pK^{\circ}_{s0} = 9.0 \ (K^{\circ}_{s0} \text{ units are } mol^2\ dm^{-6}).$$

The mean ionic activity coefficient was calculated from the Davies equation in the form: $\log \gamma_{\pm} = -A\,[(I)^{1/2}/(1 + (I)^{1/2}) - (1/3)I]$, where I is the ionic strength in mol dm^{-3} and the value of A used was 1.900 $mol^{-1/2}dm^{3/2}$.

AUXILIARY INFORMATION

METHOD/APPARATUS/PROCEDURE:

Probably UV spectrophotometry.
No other details.

SOURCE AND PURITY OF MATERIALS:

Not stated.

ESTIMATED ERROR:

Not specified.

REFERENCES:

COMPONENTS:	ORIGINAL MEASUREMENTS:
(1) Tetraphenylarsonium tetraphenylborate (1-); $C_{48}H_{40}BAs$; [15627-12-0] (2) 1-Methyl-2-pyrrolidinone (N-Methyl-2-pyrrolidone); C_5H_9NO; [872-50-4]	Alexander, R.; Parker, A. J.; Sharp, J. H.; Waghorne, W. E. *J. Am. Chem. Soc.* 1972, *94*, 1148-58.
VARIABLES:	PREPARED BY:
One temperature: 25°C	Orest Popovych

EXPERIMENTAL VALUES:

The solubility (ion-activity) product of $Ph_4As\ BPh_4$ in N-methyl-2-pyrrolidone was reported as:

$$pK^\circ_{s0} = 3.3 \ (K^\circ_{s0} \text{ units are mol}^2 \text{ dm}^{-6}).$$

The mean ionic activity coefficient was calculated from the Davies equation in the form: $\log \gamma_\pm = -A[(I)^{1/2}/(1 + (I)^{1/2}) - (1/3)I]$, where I is the ionic strength in mol dm^{-3} and the value of A used was 2.004 $mol^{-1/2}\ dm^{3/2}$.

AUXILIARY INFORMATION

METHOD/APPARATUS/PROCEDURE:	SOURCE AND PURITY OF MATERIALS:
Probably UV spectrophotometry. No other details.	Not stated.
	ESTIMATED ERROR:
	Not specified.
	REFERENCES:

COMPONENTS:

(1) Tetraphenylarsonium tetraphenylborate (1-); $C_{48}H_{40}BAs$; [15627-12-0]

(2) Nitromethane; CH_3NO_2; [75-52-5]

ORIGINAL MEASUREMENTS:

Alexander, R.; Parker, A. J. Sharp; J. H.; Waghorne, W. E. *J. Am. Chem. Soc.* 1972, *94*, 1148-58.

VARIABLES:

One temperature: 25°C

PREPARED BY:

Orest Popovych

EXPERIMENTAL VALUES:

The solubility (ion-activity) product of $Ph_4As\ BPh_4$ in nitromethane was reported as:

$$pK^\circ_{s0} = 5.7\ (K^\circ_{s0} \text{ units are } mol^2\ dm^{-6}).$$

The mean ionic activity coefficient was calculated from the Davies equation in the form: $\log \gamma_{\pm} = -A\ [(I)^{1/2}/(1 + (I)^{1/2}) - (1/3)I]$, where $\underline{I}$ is the ionic strength in mol dm^{-3} and the value of $\underline{A}$ used was 1.479 $mol^{-1/2}\ dm^{3/2}$.

AUXILIARY INFORMATION

METHOD/APPARATUS/PROCEDURE:

Probably UV spectrophotometry. No other details.

SOURCE AND PURITY OF MATERIALS:

Not stated.

ESTIMATED ERROR:

Not specified.

REFERENCES:

COMPONENTS:

(1) Tetraphenylarsonium tetraphenylborate (1-); $C_{48}H_{40}BAs$; [15627-12-0]

(2) Propanediol-1,2-carbonate (propylene carbonate); $C_4H_6O_3$; [108-32-7]

ORIGINAL MEASUREMENTS:

Alexander, R.; Parker, A. J. Sharp; J. H.; Waghorne, W. E. *J. Am. Chem. Soc.* 1972, *94*, 1148-58.

VARIABLES:

One temperature: 25°C

PREPARED BY:

Orest Popovych

EXPERIMENTAL VALUES:

The solubility (ion-activity) product of $Ph_4As\ BPh_4$ in propylene carbonate was reported as:

$$pK^\circ_{s0} = 4.6 \ (K^\circ_{s0} \text{ units are mol}^2 \text{ dm}^{-6}).$$

The mean ionic activity coefficient was calculated from the Davies equation in the form: $\log \gamma_\pm = -A\ [(I)^{1/2}/(1 + (I)^{1/2} - (1/3)I]$, where $\underline{I}$ is the ionic strength in mol dm^{-3} and the value of $\underline{A}$ used was 0.661 $mol^{-1/2}\ dm^{3/2}$.

AUXILIARY INFORMATION

METHOD/APPARATUS/PROCEDURE:

Probably UV spectrophotometry. No other details.

SOURCE AND PURITY OF MATERIALS:

Not stated.

ESTIMATED ERROR:

Not specified.

REFERENCES:

COMPONENTS:

(1) Tetraphenylarsonium tetraphenylborate (1-); $C_{48}H_{40}BAs$; [15627-12-0]

(2) 2-Propanone (acetone); C_3H_6O; [67-64-1]

ORIGINAL MEASUREMENTS:

Alexander, R.; Parker, A. J.; Sharp, J. H.; Waghorne, W. E. *J. Am. Chem. Soc.* 1972, *94*, 1148-58.

VARIABLES:

One temperature: 25°C

PREPARED BY:

Orest Popovych

EXPERIMENTAL VALUES:

The solubility (ion-activity) product of $Ph_4As\ BPh_4$ in acetone was reported as:

$$pK^{\circ}_{s0} = 8.0 \ (K^{\circ}_{s0} \text{ units are mol}^2 \text{ dm}^{-6}).$$

The mean ionic activity coefficient was calculated from the Davies equation in the form: $\log \gamma_{\pm} = -A\ [(I)^{1/2}/(1 + (I)^{1/2}) - (1/3)I]$, where I is the ionic strength in mol dm^{-3} and the value of A used was 3.760 $mol^{-1/2}\ dm^{3/2}$. The solubility products and ionic strengths were "adjusted to infinite dilution by iteration, to allow for incomplete dissociation..."

AUXILIARY INFORMATION

METHOD/APPARATUS/PROCEDURE:

Probably UV spectrophotometry. No other details.

SOURCE AND PURITY OF MATERIALS:

Not stated.

ESTIMATED ERROR:

Not specified.

REFERENCES:

COMPONENTS:	EVALUATOR:
(1) Tetraphenylarsonium tetraphenylborate (1-); $C_{48}H_{40}BAs$; [15627-12-0] (2) Tetrahydrothiophene-1,1-dioxide (sulfolane, tetramethylene sulfone); $C_4H_8O_2S$; [126-33-0]	Orest Popovych, Department of Chemistry, City University of New York, Brooklyn College, Brooklyn, N. Y. 11210, U. S. A. November 1979

CRITICAL EVALUATION:

There are two data pertaining to the solubility of tetraphenylarsonium tetraphenylborate (Ph_4AsBPh_4) in sulfolane, both determined in Parker's laboratory at 303 K (1,2). It is the first datum, the concentration solubility product in $mol^2\ dm^{-6}$, expressed as $pK_{s0} = 5.0$, that provides us with the most reliable means of calculating the solubility. Taking the solubility as $(K_{s0})^{\frac{1}{2}}$ and using the precision of ±0.1 pK units as estimated by the authors, we obtain: $(3.2 \pm 0.4) \times 10^{-3}\ mol\ dm^{-3}$. Considering the lack of information on the temperature control and the fact that the solubility estimate has only two significant figures, the above value can be considered no better than tentative.

The thermodynamic solubility product reported later as $pK^{\circ}_{s0} = 5.2$ (2) is less reliable for calculating the solubility, as the degree of the activity correction incorporated in it is not readily known.

References:

1. Parker, A. J.; Alexander, R. *J. Am. Chem. Soc.* 1968, *90*, 3313.
2. Alexander, R.; Parker, A. J.; Sharp, J. H. Waghorne, W. E. *J. Am. Chem. Soc.* 1972, *94*, 1148.

COMPONENTS:

(1) Tetraphenylarsonium tetraphenylborate (1−); $C_{48}H_{40}BAs$; [15627-12-0]

(2) Tetrahydrothiophene-1,1-dioxide (sulfolane, tetramethylene sulfone); $C_4H_8O_2S$; [126-33-0]

ORIGINAL MEASUREMENTS:

Parker, A. J.; Alexander, R. *J. Am. Chem. Soc.* 1968, *90*, 3313-9.

VARIABLES:

One temperature: 30°C

PREPARED BY:

Orest Popovych

EXPERIMENTAL VALUES:

The formal (concentration) solubility product of $Ph_4As\ BPh_4$ in sulfolane was reported as:

$$pK_{s0} = 5.0\ (K_{s0} \text{ units are mol}^2\ \text{dm}^{-6}).$$

AUXILIARY INFORMATION

METHOD/APPARATUS/PROCEDURE:

UV spectrophotometry on solutions saturated under nitrogen, using a Unicam SP500 spectrophotometer. Saturated solutions were prepared by shaking for 24 hours at 35°C and then for a further 24 hours at 30°C.

SOURCE AND PURITY OF MATERIALS:

The purification of materials has been described in the literature (1-3).

ESTIMATED ERROR:

Absolute precision was estimated to be ±0.1 pK units.

REFERENCES:

1) Clare, B. W.; Cook, D.; Ko, E. C. F.; Mac, Y. C.; Parker, A. J. *J. Am. Chem. Soc.* 1966, *88*, 1911.
2) Alexander, R.; Ko, E. C. F.; Mac, Y. C.; Parker, A. J. *J. Am. Chem. Soc.* 1967, *89*, 3703.
3) Parker, A. J. *J. Chem. Soc. A* 1966, 220.

COMPONENTS:

(1) Tetraphenylarsonium tetraphenylborate (1-); $C_{48}H_{40}BAs$; [15627-12-0]

(2) Tetrahydrothiophene-1,1-dioxide (sulfolane, tetramethylene sulfone); $C_4H_8O_2S$; [126-33-0]

ORIGINAL MEASUREMENTS:

Alexander, R.; Parker, A. J.; Sharp, J. H.; Waghorne, W. E. *J. Am. Chem. Soc.* 1972, *94*, 1148-58.

VARIABLES:

One temperature: 30°C

PREPARED BY:

Orest Popovych

EXPERIMENTAL VALUES:

The solubility (ion-activity) product of $Ph_4As\ BPh_4$ in sulfolane was reported as:

$$pK^{\circ}_{s0} = 5.2 \text{ } (K^{\circ}_{s0} \text{ units are mol}^2 \text{ dm}^{-6}).$$

The mean ionic activity coefficient was calculated from the Davies equation in the form: $\log \gamma_{\pm} = -A\ [(I)^{1/2}/(1 + (I)^{1/2}) - (1/3)I]$, where $\underline{I}$ is the ionic strength in mol dm^{-3} and the value of $\underline{A}$ used was 1.244 $mol^{-1/2}\ dm^{3/2}$.

AUXILIARY INFORMATION

METHOD/APPARATUS/PROCEDURE:

Probably UV spectrophotometry.
No other details.

SOURCE AND PURITY OF MATERIALS:

Not stated.

ESTIMATED ERROR:

Not specified.

REFERENCES:

COMPONENTS:

(1) Tetraphenylphosphonium tetraphenylborate (1-); $C_{48}H_{40}BP$; [15525-15-2]

(2) Water; H_2O; [7732-18-5]

EVALUATOR:
Orest Popovych, Department of Chemistry, City University of New York, Brooklyn College, Brooklyn, N. Y. 11210, U. S. A.

December 1979

CRITICAL EVALUATION:

The very low solubility of tetraphenylphosphonium tetraphenylborate ($Ph_4P\ BPh_4$) in water would render its direct determination unreliable, and it has never been reported. However, its solubility (ion-activity) product in water was evaluated indirectly (1) using the experimentally determined solubility product of $Ph_4P\ BPh_4$ in acetonitrile and the transfer activity coefficient of $Ph_4P\ BPh_4$ in acetonitrile calculated from those of other electrolytes. Under the circumstances, the calculated pK°_{s0} of $Ph_4P\ BPh_4$ in water and the solubility value derived from it represent data worthy of evaluation and application.

The above calculation made use of the fact that the transfer activity coefficient of $Ph_4P\ BPh_4$ in acetonitrile (i.e., for the transfer from water to acetonitrile) is related to the solubility products in the two solvents as follows:

$$\log {}_m\gamma_\pm^2 = pK^\circ_{s0}\ (\text{acetonitrile}) - pK^\circ_{s0}\ (\text{water}) \qquad (1)$$

where $\log {}_m\gamma_\pm^2$ is the transfer activity coefficient, which in this evaluation is expressed on the weight basis (molal scale), as are the ion-activity products K°_{s0}. Since the pK°_{s0} for $Ph_4P\ BPh_4$ in acetonitrile was reported as 5.68 ± 0.05 (1) (see compilation), the corresponding pK°_{s0} in water could be calculated from Equation 1 if the value of $\log {}_m\gamma_\pm^2$ were known. The latter was calculated from the relationship:

$$\log {}_m\gamma_\pm^2(Ph_4P\ BPh_4) = \log {}_m\gamma_\pm^2(Ph_4P\ Pi) + \log {}_m\gamma_\pm^2(KBPh_4) - \log {}_m\gamma_\pm^2(KPi) \quad (2)$$

Substituting into Equation 2 the corresponding experimentally determined values (1), where Pi is the picrate ion, we obtain:

$$\log {}_m\gamma_\pm^2(Ph_4P\ BPh_4) = (-6.15 \pm 0.04) + (-4.68 \pm 0.04) - (0.62 \pm 0.04) = -11.45 \pm 0.07.$$

Consequently, from Equation 1:

$$pK^\circ_{s0}(\text{water}) = (5.68 \pm 0.05) - (-11.45 \pm 0.07) = 17.13 \pm 0.09.$$

The solubility of $Ph_4P\ BPh_4$ in water taken simply as $(K^\circ_{s0})^{1/2}$ is therefore: $(2.7 \pm 0.3) \times 10^{-9}$ mol kg^{-1}. Of course, the solubility value would be the same in units of mol dm^{-3} as well. This values of the solubility and the solubility product should be considered tentative.

REFERENCES:

(1) Popovych, O; Gibofsky, A.; Berne, D. H. *Anal. Chem.* 1972, *44*, 811.

COMPONENTS:

(1) Tetraphenylphosphonium tetraphenylborate (1-); $C_{48}H_{40}BP$; [15525-15-2]

(2) Acetonitrile; C_2H_3N; [75-05-8]

ORIGINAL MEASUREMENTS:

Popovych, O.; Gibofsky, A.; Berne, D. H. *Anal. Chem.* 1972, *44*, 811-7.

VARIABLES:

One temperature: 25.00°C

PREPARED BY:

Orest Popovych

EXPERIMENTAL VALUES:

The solubility (ion-activity) product of tetraphenylphosphonium tetraphenylborate ($Ph_4P\ BPh_4$) in acetonitrile was reported as:

$$pK^{\circ}_{s0} = 5.68 \quad (K^{\circ}_{s0} \text{ units are mol}^2\ \text{kg}^{-2}).$$

The mean molar activity coefficient was calculated using the relationship:

$$-\log y_{\pm} = \frac{1.64\ C^{1/2}}{1 + 0.485\ \mathring{a}C^{1/2}}$$

where C was the solubility in mol dm^{-3} and å, the mean ion-size parameter. Although the article states that an ion size of 0.5 nm was adopted for the tetraphenyl ions, this is an error. According to the calculations in the research notebook of A. Gibofsky (1), the magnitude of å used in the calculations was 0.82 nm. The actual values for the solubility and the activity coefficient were not published, but they were 1.262×10^{-3} mol dm^{-3} and $y_{\pm} = 0.889$, respectively (1). The resulting $pK^{\circ}_{s0} = 5.90$ (K_{s0} units of $mol^2\ dm^{-6}$). The published value of $pK^{\circ}_{s0} = 5.68$ was calculated from the above using the density of acetonitrile, 0.777 g cm^{-3}.

AUXILIARY INFORMATION

METHOD/APPARATUS/PROCEDURE:

Ultraviolet spectrophotometry using a Cary Model 14 spectrophotometer. Saturation achieved by shaking the salt suspension for 2 weeks in water-jacketed flasks. Solutions filtered and analyzed at 266 and 274 nm. All solutions and containers were deaerated.

SOURCE AND PURITY OF MATERIALS:

Acetonitrile (Matheson, spectroquality) was refluxed for 24 hrs over CaH_2 and fractionally distilled. $Ph_4P\ BPh_4$ was prepared from the chloride (Alfa Inorganics, Inc.) and $NaBPh_4$ (Fisher, 99.7%). It was recrystallized three times from 3:1 acetone-water and dried *in vacuo* at 80°C.

ESTIMATED ERROR:

Precision ±10% in K°_{s0} (authors)
Temperature control: ±0.01°C

REFERENCES:

1. Gibofsky, A., Unpublished research, Brooklyn College, 1969.

COMPONENTS:

(1) Tetraphenylphosphonium tetraphenylborate (1-); $C_{48}H_{40}BP$; [15525-15-2]

(2) 1,1-Dichloroethane; $C_2H_4Cl_2$; [75-34-3]

ORIGINAL MEASUREMENTS:

Abraham. M. H.; Danil de Namor, A. F. *J. Chem. Soc. Faraday Trans. 1* 1976, *72*, 955-62.

VARIABLES:

One temperature: 25°C

PREPARED BY:

Orest Popovych

EXPERIMENTAL VALUES:

The authors reported the solubility of $Ph_4P\ BPh_4$ in 1,1-dichloroethane as:
3.22×10^{-4} mol dm^{-3}.
Using an estimated association constant of 3.60×10^3 mol^{-1} dm^3 and an ion-size parameter of $\mathring{a} = 0.66$ nm with which to calculate the mean ionic activity coefficient from the extended Debye-Hückel equation, they obtained for the standard Gibbs free energy of solution:

$\Delta G^\circ_s = 10.36$ kcal mol^{-1}
$= 43.37$ kJ mol^{-1} (compiler).

The solubility (ion-activity) of $Ph_4P\ BPh_4$ can be calculated from the relationship:
$\Delta G^\circ_s = -RT \ln K^\circ_{s0}$, yielding $pK^\circ_{s0} = 7.595$, where K°_{s0} units are mol^2 dm^{-6} (compiler).

AUXILIARY INFORMATION

METHOD/APPARATUS/PROCEDURE:

Evaporation and weighing. Saturated solutions prepared by shaking the suspensions for several days at 25°C. No solvate was detected. Method of temperature control was not specified.

SOURCE AND PURITY OF MATERIALS:

The solvent was shaken with anhydrous K_2CO_3, passed through a column of basic activated alumina into distillation flask and fractionated under N_2 through a 3-foot column. At least 10% of distillate was rejected, the rest collected over freshly activated molecular sieve. Source and purification of $Ph_4P\ BPh_4$ were not specified.

ESTIMATED ERROR:

Precision of 0.1 kcal mol^{-1} in ΔG°_s.

REFERENCES:

COMPONENTS:

(1) Tetraphenylphosphonium tetraphenylborate (1-); $C_{48}H_{40}BP$; [15525-15-2]

(2) 1,2-Dichloroethane; $C_2H_4Cl_2$; [107-06-2]

ORIGINAL MEASUREMENTS:

Abraham, M. H.; Danil de Namor, A. F. *J. Chem. Soc. Faraday Trans. 1* 1976, *72*, 955-62.

VARIABLES:

One temperature: 25°C

PREPARED BY:

Orest Popovych

EXPERIMENTAL VALUES:

The authors reported the solubility of $Ph_4P\ BPh_4$ in 1,2-dichloroethane as:
4.87×10^{-3} mol dm^{-3}.
Using an estimated association constant of 6.00×10^2 mol^{-1} dm^3 and an ion-size parameter of $\mathring{a}$ = 0.66 nm with which to calculate the mean ionic activity coefficient from the extended Debye-Hückel equation, they obtained for the standard Gibbs free energy of solution:
ΔG_s° = 7.91 kcal mol^{-1}
- 33.1 kJ mol^{-1} (compiler).
The solubility (ion-activity) product of $Ph_4P\ BPh_4$ can be calculated from the relationship:
$\Delta G_s^\circ = -RT \ln K_{s0}^\circ$, yielding pK_{s0}° = 5.799, where K_{s0}° units are mol^2 dm^{-6} (compiler).

AUXILIARY INFORMATION

METHOD/APPARATUS/PROCEDURE:

Evaporation and weighing. Saturated solutions prepared by shaking the suspensions prepared for several days at 25°C. No solvate was detected. Method of temperature control was not specified.

SOURCE AND PURITY OF MATERIALS:

The solvent was shaken with anhydrous K_2CO_3, passed through a column of basic activated alumina into distillation flask and fractionated under N through a 3-foot column. At least 10% of distillate was rejected, the rest collected over freshly activated molecular sieve. Source and purification of $Ph_4P\ BPh_4$ were not specified.

ESTIMATED ERROR:

Precision of 0.1 kcal mol^{-1} in ΔG_s°.

REFERENCES:

SYSTEM INDEX

Underlined page numbers refer to evaluation text and those not underlined to compiled tables. All compounds, except tetraphenylborates are listed as in Chemical Abstract indexes. For example, toluene is listed as benzene, methyl-, and dimethylsulfoxide is listed as methane, sulfinylbis-. The second and subsequent components of ternary and multicomponent systems are given as molecular formulae rather than chemical name. In this volume only tetraphenylborates are listed under their IUPAC name.

REGISTRY NUMBER INDEX

Underlined page numbers refer to evaluation text and those not underlined to compiled tables.

Underlined page numbers refer to evaluation text and those not underlined to compiled tables.